THE HIIT BIBLE

THE HIIT BIBLE

HIGH-INTENSITY INTERVAL TRAINING

SUPERCHARGE YOUR BODY AND BRAIN

STEVE BARRETT

B L O O M S B U R Y

LONDON · OXFORD · NEW YORK · NEW DELHI · SYDNEY

Bloomsbury Sport
An imprint of Bloomsbury Publishing Plc

50 Bedford Square   1385 Broadway
 London   New York
WC1B 3DP   NY 10018
  UK   USA

www.bloomsbury.com

BLOOMSBURY and the Diana logo are trademarks of Bloomsbury Publishing Plc

First published 2017
© Steve Barrett, 2017
Photos © Escape Creative and Getty Images pp. 192, 195 and 198

British Library Cataloguing-in-Publication Data
A catalogue record for this book is available from the British Library.

Library of Congress Cataloguing-in-Publication data has been applied for.

Print: 978-1-4729-3219-8
Epub: 978-1-4729-3218-1
EPDF: 978-1-4729-3221-1

2 4 6 8 10 9 7 5 3 1

Bloomsbury Publishing Plc makes every effort to ensure that the papers used in the manufacture of our books are natural, recyclable products made from wood grown in well-managed forests. Our manufacturing processes conform to the environmental regulations of the country of origin.

To find out more about our authors and books visit www.bloomsbury.com. Here you will find extracts, author interviews, details of forthcoming events and the option to sign up for our newsletters.

Printed and bound in China by C&C Offset Printing Co

TABLE OF CONTENTS

PART ONE: STARTING OUT   7

Safety briefing   8

Introduction   15

The Return On Investments (ROI) from HIIT   20

The science of HIIT   28

The brand names   32

Learn it, then work it!   59

PART TWO: WHAT YOU NEED   63

HIIT kit   64

HIIT fuel and fluids   82

PART THREE: THE EXERCISES   95

Bodyweight HIIT exercises   96

Bodyweight HIIT exercises intensified with dumbbells   118

HIIT exercises for the gym, garage or Crossfit™ box   140

Cardio sessions on machines   152

PART FOUR: THE WORKOUTS   167

The only HIIT exercises you'll ever need   168

Bodyweight moves   173

Gym-based moves   183

Outdoor moves   193

PART FIVE: HIIT ASSESSMENT AND TRAINING CHALLENGES   201

HIIT Assessment   202

HIIT Training Challenge   205

14-Day 'Zero to Hero' Challenge   208

28-Day 'Zero to Super Hero' Challenge   214

PART SIX: QUICK-HIIT ANSWERS   227

Author biography   236

Index   238

PART ONE

STARTING OUT

SAFETY BRIEFING

Starting a book with a safety briefing is a bit like going on a first date and telling the person all your 'issues' before saying 'hello', but by working out safely you will enjoy the amazing benefits of HIIT and avoid becoming susceptible to any injury.

First off, it is important to say that exercise has more positive effects than it does potential negative ones. People who start a progressive exercise regime with a body that is free from disease, injury or physical restriction will feel better than they did when they did nothing. A fact of exercise, however, is that some people suffer injuries, the cause of which can sometimes be difficult to establish because often there are contributing factors than have a knock-on effect. These may include what a person has been doing in the hours preceding their workout, or what they have or haven't eaten or drunk; being dehydrated, for instance, has a detrimental effect upon a subsequent exercise session and therefore may be a factor if injury subsequently ensues.

So, my approach is this: I am responsible for the
contents of this book. However, conveying information
via the written word is not quite the same as delivering
personal training, coaching or presenting in the flesh – being
somewhat conceptual rather than prescriptive – so we all have
to accept that there are limitations to how literally that information
can be treated. Moreover, different people have different perceptions
of what's easy/hard, simple/complex or achievable/unrealistic so you need
to use your own judgement. As a general rule of thumb, though, I recommend
that you always use caution rather than being reckless.

This advice may seem counter-intuitive in a book that is all about a method of
exercise that requires you to push yourself to the upper edge of your physical limits,
but in fact it tallies with my view that you will achieve more from the impact of
exercise by being smart rather than simply 'smashing' yourself during every
session. Preparation, listening to your body and understanding your limits is key
to a rewarding and injury-free workout.

I have been doing HIIT for more than 30 years without suffering any ill-effects
or injuries. Sure, I've had plenty of days after workouts when I'm forced to
walk down the stairs in slow motion or struggle when putting on a T-shirt
because the muscles in my chest, arms and shoulders are screaming for
forgiveness, but these are all short-term niggles that are the normal
consequence of challenging the body to function and perform at a higher
level. If you can't cope with DOMS (Delayed Onset Muscle Soreness),
getting soaked in your own sweat or the feeling that your heart and lungs
are operating at the peak of their ability, then HIIT is not for you – and the chances
are that you will misinterpret these signs that you've done a great workout as being
an injury. Muscle discomfort after exercise (typically the day after) is to be expected;
in fact, many fitness professionals describe the aftermath of a good workout as 'God's
tax on exercise', and it will pass.

While creating the exercises and workout programmes in this book, I've thought
about every exercise long and hard and tried to present them in such a way that they

are easy to understand and perform. This style is, I hope, best described as 'clear and concise', conveying in written form a skill I have been told by clients I have for 'making really complicated stuff easy to understand'.

The information contained within this book is based on current research and the recommendations of governments and health organisations, and is safe for adults over 18 years of age. With this in mind, note that the exercises and workout programmes are also intended to be effective – and are therefore challenging even for experienced fitness enthusiasts.

To be effective, as with all types of exercise, HIIT involves an element of risk. In light of this, please read the following passage:

You should consult your doctor or other health-care professional before starting this or any other fitness programme to establish if it is appropriate for your needs. This is particularly true if you (or your family) have a history of any condition related to cardiac respiratory function (high blood pressure, heart disease, asthma, angina, etc.) or if you have ever experienced chest pain. If you smoke, High-Intensity Interval Training (HIIT) will not only be less effective but potentially impossible. If you have high cholesterol or are obese, then seek advice and approval (most doctors will be pleased that you are taking action, but do check first). Likewise, if you have a bone or joint problem that could be made worse by a change in physical activity, you should also check first with a medical practitioner. Nothing in this book replaces or overrules the advice of a medically qualified doctor, so do not start this fitness programme if your doctor or other health-care professional advises against it.

'Listen' to your body and if you experience faintness, dizziness, pain or shortness of breath at any time while exercising, stop immediately.

This book offers health and fitness advice relating to HIIT and is designed for information/inspirational purposes only. Do not disregard, avoid or delay obtaining medical or health-related advice from your health-care professional because of something you may have read in this book. The use of any information provided in this book is entirely at your own risk.

Furthermore, subsequent developments in medical/health-related research may impact the health, fitness and nutritional advice within this book.

The images contained in the book show people doing exercise in suitable attire, which may include short shorts and sports bras. All the items of clothing are standard and come from well-known global brands, and therefore should be viewed as performance-enhancing products rather than something that is intended to be in any way offensive or sexually provocative.

Phew, now let's lighten the mood. My mission is to enlighten you and make you aware of the benefits of high-intensity activity – with the goal being that you can improve your health, fitness and capacity to exercise – but it is as well to stress that your doctor knows best, so consult your general practitioner before attempting anything in this book.

HIIT health checklist

Before doing HIIT you must be able to answer NO to ALL of these questions:

☐ Do you or have you ever felt pain in your chest?

☐ Have you been told by your doctor that you should only do physical activity recommended by a medical practitioner because you have a heart condition or another relevant medical condition?

☐ Is your doctor currently prescribing you medication for a blood-pressure or heart condition?

☐ Do you lose your balance as a result of dizziness or do you ever lose consciousness/collapse?

☐ Do you have a bone or joint problem (especially of the back, knee, hip, shoulder, elbow or wrist) that could be made worse by exercise?

☐ Do you know of any other reason why you should not do physical activity?

Those six questions are 'deal breakers'. If you said yes to any of them then you really shouldn't proceed with any of the suggested exercises in this book. Please do read and enjoy the contents, but you need to get some personalised advice from a medical professional and consider a period of 'pre-conditioning'. It is important to understand, however, that you are absolutely NOT a 'write-off' and exercise in all likelihood will be very beneficial, it's just that HIIT may not be the best approach.

HIIT reality check

Answer the five questions below to ensure you are ready specifically for HIIT. Intensity is measurable and therefore something that can be quantified by the reaction HIIT places on your body. These are more extreme if you are not used to working out. So if your answer to the following statements is 'yes' or 'umm, maybe' then you may need to make some lifestyle changes before you begin.

☐ Can you currently not exercise for at least 20 continuous minutes at a moderate intensity? If exercise intensity were to be graded 1–10 with 10 being the toughest, can you do 20 minutes at level 6.5-plus?

☐ Are you on a restricted-calorie diet (fewer than 2,000 calories per day for women and 2,500 calories per day for men)?

☐ Are you perpetually dehydrated (highly likely if you drink less than 1 litre (2 pints) of fluids per day)?

☐ Do you find it hard to move freely without the need to groan or feeling tight/ restricted? A lack of mobility will impair your ability to do some of the exercises in this book.

☐ Does your level of body fat affect your ability to move? This isn't a heartless question, it's simply a reality that a higher level of fat weight rather than lean-muscle weight will make the fast movements required in HIIT to be potentially harmful.

If you have answered 'yes' to some these questions, it may mean that your 'start date' with HIIT needs to be delayed. As a fitness professional, I would rather you succeed in the long term rather than dive into a new regime and prematurely crash out because you do too much too soon. HIIT is faster and more exhilarating, intense and stimulating (both chemically and mentally) than other types of exercise. If HIIT were a horse, it would be a beautiful unbroken stallion capable of doing serious damage to an inexperienced rider. If you aren't ready, don't be a hero – start with something more placid and look forward to the ride of your life once you have built up some skills, resilience and, most importantly, body awareness.

INTRODUCTION

P90X™, Tabata™, Spinning™, INSANITY®, CrossFit™, Sprint™ – these are possibly all words that will be unfamiliar to anyone not involved in fitness. For those of us who have recently ventured into a health club, gym or fitness studio anywhere in the world, though, this list will fill you with either excitement or dread because they are some of the most popular HIIT fitness programmes available. Although they all fall under the umbrella term 'HIIT', they each pose a completely different set of challenges, and result in different emotions and outcomes. What they do have in common is that they will cause you to get hot, out of breath, maybe feel pain and perhaps give you a 'buzz' that lasts long after the session is over. All of those are ordinary emotions and physical responses to the intense physical activity of HIIT.

HIIT is in essence 'just' a formula: frenetic exertion + recovery x sets = High-Intensity Interval Training. The words in the equation that have been examined, experienced and occasionally hated during the development of this book are High-Intensity and Interval.

The 'high-intensity' element represents training in short bursts at a maximum level of effort and the 'interval' 'bit' isn't either the work or the recovery periods: it's both, and they can range from many minutes to literally seconds. Moreover, I'll explain how these impact the workout and, crucially, the results.

I warn you now that there isn't a single answer to the question: 'What is the best HIIT method?' In fact, it actually seems remarkable that a multi-billion fitness industry has managed to develop without a definitive definition of what this term HIIT – which is tacked on to the names of so many classes and methods – actually is. That said, having spent 30 years in this business, this lack of consensus shouldn't surprise me, since other functional-training and core-strength practitioners are also sometimes unable to give a definitive description of what their methods achieve.

Despite this, we can at least agree on one thing: HIIT is universally acknowledged as being a highly effective way of getting people of all abilities fit (bear in mind that HIIT is even used for cardiac rehab patients, so it's certainly very 'inclusive').

In fact, HIIT is now considered a global phenomenon by most trainers, coaches and trend-watchers, and the American College of Sports Medicine (ACSM) has put HIIT at the top of their list of global trends in health and fitness. It is not, however, without its critics, and even the ACSM has expressed concern that the intense nature of HIIT may actually be putting people off exercise and damaging their perception of what they need to do to improve or maintain their health and fitness.

But let's get real here. HIIT isn't actually 'new' – the way it's being packaged might be, but the methodology has been in use for decades, so my view is that ACSM's concerns are a distraction. What they are saying is that HIIT might be so intense that it could potentially put a beginner off becoming a regular exerciser because they will look at HIIT and decide that 'if that's what exercise looks like then I can't do it'. Such issues are psychological and philosophical rather physiological, and in my opinion don't apply to those who are in good-enough shape to cope with the challenges of HIIT. I genuinely think HIIT is awesome and such concerns are outweighed by its clear benefits.

If you combine all the hype and media attention, and the fact that HIIT has actually been researched and proven to be effective, it's easy to see why it has caused a global shift in the way people work out. Now you could say or think that anything can become popular if you throw enough money at it, but HIIT has managed to retain its independence despite some fairly massive global companies trying to suggest they have ownership of the method. Sure, these organisations have ownership of their version of HIIT but, as you'll read later, if you distill the details and data sometimes the only thing that seems to set them apart is the amount of time spent working and the duration of the recovery phase.

You can HIIT … indoors, outdoors, in water, in the gym with or without equipment, on a bike, rower or treadmill, and, best of all, if you strip HIIT back to the bare bones you can work with nothing more than your own bodyweight or a pair of 5–10kg (10–20lb) dumbbells. Doing the right compound exercises all you'll need is about 2 x 2m (6.5 x 6.5ft) of space and a big slice of determination.

Moreover, in many ways HIIT is speedy not just in terms of the velocity at which the exercises are performed, but also speedy in terms of how long you need to spend doing it for in order to see results. This is a bonus for those who are time poor.

"You can HIIT …
indoors, outdoors,
in water, in the
gym with or without
equipment … "

THE RETURN ON INVESTMENTS (ROI) FROM HIIT

HIIT isn't new, in fact, 30 years ago we were already doing forms of high-intensity interval training. So if this has been around for 30 years, why is it suddenly all the rage now? Because, like so many things in the world fitness that have been around for a long time, in the old days we did them without necessarily understanding how they worked. In recent years, research has indicated the clear benefits, and backed up by this evidence, companies have been able to package it as a product and make it available across the world.

When it comes to personal fitness, generally we all prefer rapid results with small amounts of time invested. This is natural, as most people given the option would prefer to spend less time working out. This is especially the case if their wish list is to spend as little time as possible getting (and maintaining) a lean functional, fast, agile body. HIIT is hard, but the softer options just don't seem to work as effectively.

It is the effectiveness and its ROI that that makes HIIT so interesting, with it offering major fitness gains achievable with regular exposure to HIIT.

The list of gains is extensive and these can be viewed as the building blocks for long- and short-term improvements to our body function. These include effective weight loss, improved cardiovascular health, increased endurance levels, increased human growth hormone, enhanced blood quality, slows ageing, improved mood and brain function, and limits levels of cortisol.

EFFECTIVE WEIGHT LOSS

One of the key benefits of HIIT is that it can be an excellent means of weight loss when matched with an appropriated nutrition programme. A Laval University, Quebec, research study (1994) indicated that it was more effective than moderate cardio such as running or cycling.

Scientists have also identified another rather wonderful benefit, and that is that HIIT is an effective way to stimulate the physiological phenomenon called EPOC, or to give it its full title Excess Post-Exercise Oxygen Consumption (often called 'after-burn'). In simple terms EPOC is when our body keeps burning fuel at a higher rate than it would have done if you hadn't performed a high intensity workout.

Simply put, EPOC makes the body continue to absorb oxygen at a measurable higher rate following a HIIT session, which results in the stored fat being torched for hours after you have finished a workout. Now who wouldn't want that!

It's impossible to say exactly how long and how much afterburn occurs in individuals unless you are rigged up to an oxygen-mask and computer measuring the gases going in and out of your mouth. Tests have shown that you are absorbing oxygen and burning fuel at a rate of +10 per cent three hours after a workout, and maybe 4 per cent 16 hours later. This is a lot less than the statements that come out of the mouth of many group exercise instructors who'll often be heard saying that, 'doing HIIT will make you burn more calories for 24hrs after a workout', but I think it's more realistic to think that EPOC is effective at extending the benefits of exercise for 'the rest of the day'. One significant thing we can be sure of is that the EPOC effect is 100 per cent triggered by the intensity of a workout rather than the duration of it.

IMPROVED CARDIOVASCULAR FITNESS

HIIT puts a lot of demands on the body and a number of leading research studies have indicated that the high intensity exercise can lead to greater improvements to cardiovascular health when compared with moderate intensity exercise, such as jogging.

High Intensity has also been shown to increase our maximum oxygen consumption (VO^2 max). VO^2 refers to the maximum amount of oxygen a human can utilise during intense exercise. It is measured as 'millilitres of oxygen used per minute by each kilo of bodyweight'. This effectively means that the volume of oxygen a person can extract from air inhaled is optimised, and the benefits include increased oxygen intake which when delivered to your muscles, enables you to run faster for limited periods. While VO^2 Max isn't a guarantee of greatness in the world of athletic performance, it'll certainly help you get a contention.

INCREASED PHYSICAL ENDURANCE

Repeated exposure to HIIT is known to increase an athlete's endurance (which is a fact some people will find illogical because HIIT sessions are short compared to endurance events where you perform for long periods of time). Many athletes have dramatically reduced their mileage in preparation for running a marathon so that they train hard but not long. This is especially the case for those running sub three-hour races.

INCREASED LEVELS OF HUMAN GROWTH HORMONE

Human Growth Hormone (HGH) has been described as the 'Juice of Youth'. HGH is produced in the pituitary gland located under our brain. Its purpose is to maintain or optimise the 'normal' function of structures in the body (muscles and other soft tissues) and manage metabolism. Its secretion into the bloodstream also helps keep blood glucose levels within desirable levels. HGH has been shown to increase ten-fold during HIIT sessions, which is an excellent benefit when you consider it is partly responsible for the development of lean muscle tissue.

Human growth hormone is a vital component of the human endocrine system. In childhood and adolescence, the purpose of HGH is to promote growth in height and physical stature. When you become an adult its role switches to that of being a factor in maintaining a lean muscular body mass, strengthening bones and stimulating hair and nail growth. In fact, HGH is one of the significant factors in the creation of both the feeling of internal wellness and all the things that we cosmetically associate with looking healthy.

From the age of 30 the production of our HGH decreases, this is called a somatopause which is a natural process where HGH and some of the hormones that influence youthfulness become more scarce in our system. The effects of HIIT upon the manufacture of HGH in the body are both long and short term. Low-intensity activities don't have anywhere near the same effect, as HIIT and HGH levels have been seen to increase in the bloodstream by over 700 per cent during HIIT workouts. Why? Well, it's all about the muscles that are worked during HIIT or to be more accurate the individual muscle fibres, these are the fast-twitch fibres.

This is because multiple studies have found that a high-intensity burst of activity where the heart rate beats above our anaerobic threshold (best established by VO^2 max testing) for over 20 seconds repeatedly, five or more times in one workout, the fast-twitch skeletal muscles fibres release HGH at a level never naturally seen in sedentary individuals.

Working in partnership with this, simply getting approximately 8 hours' sleep per night can also help optimise production of HGH (personally I find tough workouts and great sleep go hand in hand). Just try to start those eight hours as close to ten o'clock as you can rather than gone midnight, as this will help you also get extra benefits to your circadian rhythm (our natural caveman sleep patterns). Another way to maximise HGH gains is to avoid sugar after workouts. Consuming sugar as liquid, solid, natural or artificial and especially fructose within 2 hours of a workout will cause your hypothalamus in your brain to release somatostatin, which will decrease your production of HGH. Sugars also spike insulin levels, which increases levels of fat storage.

Just a very quick personal note on HGH, there is huge difference between the HGH you produce yourself and any that is introduced to your body via pills, needle or creams (unless it's medical grade and comes from your GP). You should never buy, swallow, inject or rub on anything that you couldn't buy at your local healthfood store. HIIT will turn your body into its very own laboratory that produces more beneficial hormones and chemicals than any other form of exercise.

ENHANCED BLOOD QUALITY

There are a number of HIIT induced outcomes which positively impact blood. These include an increased ability to absorb oxygen and the ability to shift your anaerobic threshold (the point where your body is functioning on blood that isn't being constantly replenished with 'fresh' oxygenated red stuff), a reduction of the 'bad' LDL cholesterol and increased production of the 'good' cholesterol HDL. Other benefits include an improved efficiency of the venous return system – the system for getting blood back to the heart and lungs once it has been 'used'.

IMPROVED BRAIN FUNCTION

HIIT workouts can impact the brain and its functions in a number of important ways. These include improved neuroplasticity which relates to the ability of the brains neurons and connections to rejuvenate. The rejuvenation of the brain's structure occurs only after repeated stimulation and doing HIIT creates a wash of hormones necessary to improve neuroplasticity.

In addition, recent studies of BDNF (brain-derived neurotrophic factor) production strongly suggest that it is increased by high-intensity exercise. It has been suggested that this neuropeptide hormone (the little molecules that help neurons communicate) partially contributes to the sensation known as the 'runner's high', or the sensation of a post-exercise euphoria, as unlike some of the types of endorphins manufactured in the body it can pass through the blood-brain barrier. This chemical is thought to be responsible for reinforcing the links that allows the brain to maintain its internal communication network by actually increasing the production of nerve protecting compounds.

Not only does this mean that the volume of your brain is maintained, but also its ability to function remains optimal. In short, 'washing' your brain with BDNF may stop it from shrinking in old age. So, BDNF is actively involved in both your muscles and your brain, and this cross-connection appears to be a missing link in why a physical workout can have such a beneficial impact on your mood and muscle. In addition, if you work hard enough and often enough, there is strong evidence to suggest that this can reverse brain decay and age-related muscle decay.

HIIT also helps nerve cells create proteins known as neurotrophic factors (NF). As a result of a high-intensity workout, our brains also generate this rich brain-food chemical which activates stem cells and finds its way to satellite cells (of muscles). This is then converted into new neurons and new muscle cells and, in addition, protect neuromotors from decay.

SLOWS AGEING

Telomeres have recently been discussed in relation to exercise and Dr Mercola, who is one of the leading figures in HIIT research and training, has specialised in researching the relationship between telomere decay and HIIT. His research has shown that HIIT may help preserve telomeres and this may potentially play a role in slowing the ageing process

Our body is made up of cells, and every cell has a nucleus which contains the chromosomes that in turn contain your genes. The chromosome is made up of two 'arms', and each arm contains a single molecule DNA, which is essentially a string of beads made up of units called bases.

A typical DNA molecule is about 100 million bases (layers) long, and at the tip of each arm of the chromosome you'll find the telomere. An easy way of visualising this on a massive scale is that a telomere is like the plastic tip on the end of your shoelaces.

The telomeres shorten every time your cell divides, starting at the moment of conception. If we separated the layers we would find it to be about 15,000 bases long, once your telomeres have been reduced to about 5,000 bases, you will essentially die of old age.

With the rate of erosion being pretty much constant from when we are in the womb, we can start to understand why even with a perfect super clean, antioxidant-rich diet, no one gets to live past 120 years old. The situation gets worse if you are exposed to environmental toxins, trans fats, obesity, smoking and all the other stuff we know is bad for us, as this can actually accelerate the speed at which the telomeres shorten meaning you can die of *old age* in your late 60s.

Until recently it's been believed that the telomere-shortening process could not be affected or stopped by healthy eating habits or exercise. Now, however, opinions are changing because researchers have discovered that while we can't switch off the ageing clock we may be able to slow it down.

Michael Fossel's bestselling book *The Immortality Edge* was based on Nobel prize-winning genetics research and suggests that we can keep our telomeres healthy for increased health and longevity, and HIIT forms part of this. For instance, while it hasn't been suggested that HIIT will increases the thickness or number of layers (bases), there is evidence that it can maintain the original robust nature of the tips for as long as possible.

IMPROVED MOOD

HIIT workouts increase the levels of endorphins we produce and these can have an instant effect upon your moods and energy levels, not to mention how well you cope with stressful situations.

Endorphins are biochemicals commonly referred to as neurohormones, these act by modifying the way in which nerves respond to transmitters and their effects have been regularly compared to opiate drugs (morphine and codeine). So how does all this relate to HIIT? Well, not to over simplify things, but self-inflicted 'stress' is exactly what the body experiences when we do HIIT so it's reasonable to say that the release of endorphins enables us to work hard and push through pain.

LIMITS CORTISOL

From about 2005 onwards I've often heard physical trainers tell clients that 'cardio is bad for you as it increases the stress hormone cortisol'. Too much cortisol is associated with muscle loss and the gain of belly fat – two things we want to avoid when seeking leaner bodies. Long periods of moderate intensity 'cardio' (in the gym running, rowing, cross trainer, etc.) do cause your cortisol levels to drift upwards (they do this naturally throughout the day – even if you don't exercise – until you go to sleep, at which point they start to fall until you wake up the next day, and the whole process begins over).

According to a world-renowned strength coach Charles Poliquin, cortisol only starts to creep up significantly after 20 minutes of cardio, or one hour of strength training. Given that HIIT is based on short bursts, with all but heroic super humans able to work out beyond 20 minutes at a high intensity, we can throw ourselves into HIIT without concerns about increased cortisol levels.

Fringe benefits

All of these benefits, I feel, make HIIT a worthwhile use of time compared to other forms of exercise. However, if you look at the collective benefits, it really is a remarkable form of exercise as no other single exercise protocol can deliver such an eclectic range of benefits. The other 'fringe benefits' of doing HIIT workouts is that all of the brand-named workouts and almost all of the exercises that I use are classed as functional. This means that the moves/exercises incorporate multiple moving joints, multiple muscles and also have elements of flexion, extension and rotation (frontal, sagittal and transverse to give them their technical names) in the spine as part of the movement patterns. I'm going to say that exercises that require/involve only single planes of motion at a time are unproductive but my logic is that outside of the gym environment, real life requires you to have all your muscles working as a team and the best way to develop this ability is not by working on them one at a time but rather to work them as a collection. This is best achieved not by performing single plane movements but by doing Tri-planar moves that incorporate flexion, extension and rotation of the spine.

THE SCIENCE OF HIIT

Evidence and global success in the fitness industry don't always go hand in hand; in fact some of the most familiar and profitable fitness products and programmes became global success stories as a result of big marketing budgets rather than extensive studies. That said, the evidence of the remarkable benefits of HIIT does exist. In an attempt to make sense of the science here, there is a need to impose some boundaries or criteria on what I'm going to suggest is relevant and credible. So, if I'm going to take it seriously then the process needs to be able to tick three boxes:

☐ The study was completed in the last 10 years (so we can compare like for like)

☐ The methods used in the test should be easily repeatable as a training programme (basically, I'm not interested if the only way of achieving the results is if you're high up a mountain or in some kind of facility that looks more like hospital than a gym)

☐ It's an 'everyman' solution (this is the common-sense filter that excludes any methodology that can't be integrated into the lifestyle of an average man or woman in their 20s, 30s, 40s or 50s. This means that if a specific style of HIIT leaves you so sore or exhausted that you can't carry on working or running a household, then I'm not interested)

So what exactly have the scientists agreed upon? Not much it seems, and even my usually ultra-reliable resource the ACSM is more vague about the details of HIIT than one would have expected, but at least they came to the same conclusions as me when it comes to comparing protocols. In fact, they were even more ambivalent about what's best practice, due to the impossibility of ever being able to perform a wide range of training protocols on the exact same group of people (with that group always having the exact same start point) because (and this is my assumption) every time the group followed a training regime they would change body composition/ VO_2 max and therefore end up starting the next set of tests at a higher level of personal fitness.

My reason for using ACSM's research and conclusions as my benchmark is that they have a not-for-profit approach when it comes to distilling and then publishing their guidelines. This is in contrast to some commercial organisations, which in all likelihood try to produce results that 'prove' the effectiveness of their method by using a group of unfit individuals with everything to gain, rather than a group of super-fit athletes who'll only achieve marginal gains by using the technique.

While the ACSM guidelines are of course appropriate, they are written with the mindset that athletic outcomes are always the goal, whereas the reality is that HIIT has become an industry within an industry that's interested in cosmetics and a feeling of empowerment, not just the manipulation of energy systems.

ACSM findings about HIIT

■ It's suitable for a wide range of individuals.

■ Work periods can range from 5 seconds to 8 minutes and can/should be performed at 80–90 per cent of a person's estimated maximum heart rate (EMHR).

■ The recovery periods could/should be performed at 40–50 per cent EMHR. (I disagree with this one, because whenever I've done Tabata™ splits I do nothing during the 10 seconds of prescribed recovery time. Yes my heart rate is most likely dropping to 40–50 per cent EMHR, but I'm standing still not performing.)

■ The recovery period should be much longer than is allowed in most gyms. ACSM says: 'Long interval protocols involve work/recovery time to be at a 1:1 ratio 1–5 mins work followed by 1–5 mins recovery.'

■ Another popular HIIT protocol is called the 'spring interval training method'. With this type of programme, the exerciser does about 30 seconds of 'sprint or near full-out effort', followed by 4–4½ minutes of recovery. This combination of exercise can be repeated three to five times.

Biological factors that impact on the body's energy systems

For the record, to maximise the effect upon energy systems the work/rest ratios of HIIT are based on the following factors:

Adenosine triphosphate (ATP) and phosphocreatine (PC) system

Usually referred to as the ATP-PC system, this is also sometimes called the phosphogen system. It is instant and functions without oxygen, and allows for up to approximately 12 seconds of maximum effort. During the first few seconds of any activity, stored ATP supplies the energy. For a few more seconds beyond that, PC cushions the decline of ATP until everything flips to another energy system.

Magic numbers

10–12 seconds' work; 30–60 seconds' recovery; 6–10 rounds/sets.

Anaerobic system (ATP)

This is also known as anaerobic glycolysis due to the initial process being the same as aerobic glycolysis but, as the name implies, it occurs without oxygen. Obviously, we are still respiring, but our muscles don't require a steady flow of oxygenated blood in order to function. Anaerobic exercise is physically much harder than aerobic exercise and is the energy system most likely to be functioning during HIIT session.

Magic numbers

1 minute's work; 2 minutes' recovery; 4–8 rounds/sets.

Aerobic system

The building block of creatine phosphate (CP) is now in the mix to re-synthesise the ATP. This most basic type of exercise is low intensity and requires oxygen, and wouldn't be classed as being high intensity. Aerobic activity uses fat as well as a small amount of glycogen for fuel, but the chances of achieving such a great 'buzz' as you do after/during HIIT is unlikely, as is the potential for anything that resembles free gains from EPOC or Excess Post-Exercise Oxygen Consumption (often called 'after-burn'). EPOC makes the body continue to absorb oxygen at a measurable higher rate following a HIIT session which results in the stored fat being torched for hours after you have finished a workout.

Magic numbers

3 minutes' work; 3 minutes' recovery; 6–12 rounds/sets.

THE BRAND NAMES

The HIIT industry has gone from zero to hero in less than five years, and looks like it will continue to grow (which means 'stealing' the market share from other types of activity, as well as attracting more new participants). The brands that have sprung up can be carved up into three very distinct groups: venues, methods and licensed programmes.

On the surface, there is a simmering rivalry and a healthy respect between the competing brands. Victory, you would think, would be demonstrated by a combination of positive results in the testing lab and/or people power (commercial success). In reality, the industry is as much about creating hype as it is about needing to provide any evidence, other than perhaps some fantastic before and after pictures of participants. For this reason, choosing a programme from the countless brands can be something of a minefield, but here are a few that I would say have either already proved beneficial or are tipped for greatness:

Tabata™ (see p.34) – method/licensed programme
INSANITY® (see p.36) – method/licensed programme
P90X™ (see p.42) – licensed programme
The Little Method (see p.44) – method
Les Mills Grit™ (see p.46) – licensed programme
Turbulence Training (see p.48) – method
Sprint 8® (see p.50) – method
SoulCycle™ (see p.54) – venue/method
Orangetheory Fitness™ (see p.55) – venue/method/
licensed programme
Barry's Bootcamp™ (see p.57) – venue/method
CrossFit™ (see p.40) – venue/method/licensed programme

In the lab, the leaders of the pack are Tabata™ (see
p. 34), which was created in 1996 by Professor Izumi
Tabata in Japan; and Sprint 8™, created by Phil Campbell
and Dr Mercola in the USA. Both systems were developed via
research programmes before they were launched to the public.

Then there are the venues – boutique gyms such as Barry's
Bootcamp™, SoulCycle™ and Orangetheory Fitness™. I doubt
many new customers ask for a copy of the white paper that proves
their method's effectiveness, because these companies are more about the
atmosphere, instructor and wrap-around features than they are about demonstrating
biological outcomes.

The home-workout methods of INSANITY®, P90X™, The Little Method, Les Mills Grit™
and Turbulence Training, meanwhile, have either developed organically over time
or been fast-tracked to success via boardroom investment.

Categorising CrossFit™ is something of a challenge (a nice one). This is because while
it is a venue, method and programme (which requires the payment of a licence fee), it's
also to some people a way of life that perhaps borders on being an obsession.

Tabata™

This method could so easily have ended up being called Izumi, since this is the first name of its creator, Professor Tabata. I'm pretty certain that back in the 1990s when the Japanese scientist and his colleagues started testing his theories on the Japanese Olympic speed-skating team he could never have imagined that 20 years later the rights to his method would be of interest to Hollywood, but that's exactly what happened. The reason Tabata™ has become such a household name in the world of fitness today is in no small part to do with Universal Pictures and Big Shot Productions investing in the method.

I digress. Back to the '90s, when researchers pitted two workouts against each other, using students at Ritsumeikan University Graduate School of Sport and Health Science as subjects (the ice-skating connection is a distraction used to lend authority to the

findings; most of the activity took place in a lab using a Monarch Ergometer bike rather than in an ice rink and using skates). During the tests, group one (the control group) cycled at 70rpm and at 70 per cent of their VO_2 max for one hour. Group two (the Tabata™ Protocol group) cycled hard for 20 seconds at 170 per cent of their VO^2 max, then took a 10-second rest before repeating the effort/rest cycle for a total of eight rounds. Both groups performed their sessions five times per week for six weeks. The results showed that although group 1 had increased their aerobic systems, their anaerobic systems remained the same. Group 2, however, had not only increased their aerobic systems, but they showed a 28 per cent increase in their anaerobic systems, too.

In response to the findings, Professor Tabata then carried out a second round of experiments to compare his '20 seconds on 10 seconds off' (Tabata™ Protocol) method against what happened when subjects went up another HIIT protocol to 30 seconds on with 120 seconds' recovery. Apparently, both methods delivered improvements, but the Tabata™ Protocol gains were measurable after just one week.

So, what happened next? Well not much actually, until Universal Pictures became involved in 2013 and the phrase 'Four-minute fitness, scientifically proven' was born. After this, the programme migrated from an exercise bike to a bodyweight studio group-exercise format, and the PR team came up with a way of insulting every PT and instructor who ever wrote a training programme, with their statement: 'This is the first fitness system born in the lab. It hasn't been made up by a fitness instructor or dancer, it's the result of an internationally renowned scientists clinical findings.' As the method boomed, some awesome videos were released showing beautiful people doing the Tabata™ workouts, which morphed into a combination of functional circuit moves blended with martial arts-style Capoeira primal movement patterns. (Primal is terminology for any style of exercise that encourages the body to move naturally rather than creating an artificial environment – running, jumping, squatting or throwing are all primal, while cycling or using seated weights machines aren't.)

Personally, though, I have never been able to easily achieve the levels of intensity required for the method to work using the Capoeria-style moves, whereas on a studio spin bike I can do this almost instantly. To sum up, then, Tabata™ Protocol is a good method, but it generally works better on a spin bike or if you use simple hardcore moves such as the ones outlined in this book.

INSANITY®

If you happen to turn on your TV late at night or early in the morning and flick through the channels, the chances are that it won't be long before you have Shaun T. on your screen bellowing at a group of sweaty people in a space that looks like my old school gymnasium. You have found the Beachbody Workout infomercial for the INSANITY® workout, the headlines for which include: 'I'll get you a year's worth of results in just 60 days'; 'Are you crazy enough for INSANITY?'; and 'The hardest workout ever put on DVD is now the #1 workout in America'.

The entire approach is summed up with this following statement from Shaun T. on the homepage of the INSANITY® website:

'When I created the INSANITY workout, I knew it would produce insane results in 60 days, but I wasn't sure if anyone was brave enough to try it. Turns out, there are a lot of crazy people out there. Crazy enough to actually enjoy doing the world's most insanely tough workout. To like the feeling of being drenched in sweat, of going balls-to-the-wall for a full 45 minutes of muscle-searing exercise. Is INSANITY® hard? Oh, yeah. It really IS the hardest workout ever put on DVD. It's totally crazy but it's going to get you crazy-good results.'

What I love about that statement is it says to people you would have to be crazy to try this, but if you do you'll get unreal results. This promise is backed up by testimonials from people asserting that they had previously tried 'everything' and that INSANITY® was the only thing that produced results. To compel people to post 'before' and 'after' shots that demonstrate the effectiveness of the programme, INSANITY® has come up with a clever marketing strategy that means participants only receive the much-coveted INSANITY® T-shirt bearing the logo 'you can't buy this T-shirt you have to earn it' if they submit these torso shots for scrutiny by the INSANITY® team.

The best way to describe the INSANITY® method, they say, is to compare it to a cycle class, during which you work then recover. In INSANITY®, however, instead of recovering for a couple of minutes (which is a bit unfair because it suggests a cycle class is an easy option) you only get a few seconds' recovery (time enough to 'gulp some air') in an INSANITY® session, before it's time to go again for another three to five rounds.

Is the lack of scientific verification of this method an issue? For me, the answer is no, because I've spoken to many very fit people who still say that they can't keep up with Shaun T. throughout an entire DVD. (Although of course we don't know for sure whether they actually filmed the workouts in one take or if there were some 'cuts'.) Either way, though, the workout itself is great fun and certainly gets the heart racing.

"Preparation, listening to your body and understanding your limits is key ... "

CrossFit™

Probably the fastest-growing fitness business in the world, it's not even worth me guessing how many CrossFit™ 'boxes' (they call their gyms 'boxes') there are out there because they are opening at a phenomenal rate. CrossFit™ isn't intentionally 'disruptive' as an enterprise, but their business model does have similarities with those of two of the biggest companies on the planet: Uber™ and AirBNB™. In case you don't know, Uber™ is a taxi company that neither owns any cars nor employs any drivers, and AirBNB™, without owning a single hotel, rents out more rooms per night than any of the biggest hotel companies. CrossFit™, meanwhile, runs an affiliate system whereby the 'box' owners pay a fee to be able to use the CrossFit™ logo above the door.

Despite this commercial side, CrossFit™ itself is not restricted to a specific workout or venue, and there is nothing to stop you going online and finding the WOD ('workout of the day') and doing it at home, for free. A Crossfit workout can consist of almost any credible form of exercise that fits the Crossfit model of challenging yourself on a daily basis, the mode of 'challenge' can be eclectix as a Crossfit WOD could involve the gym, the great outdoors or even a swimming pool. However, the exercise routine isn't the end of the story, and CrossFit™ sees itself as a way of way of life and a philosophy as much as a workout, as expressed by Greg Glassman, the founder and owner of CrossFit™, who sums up the fitness system thus:

'Eat meat and vegetables, nuts and seeds, some fruit, little starch and no sugar. Keep intake to levels that will support exercise but not body fat. Practice and train major lifts: Deadlift, clean, squat, presses, C&J, and snatch. Similarly, master the basics of gymnastics: pull-ups, dips, rope climb, push-ups, sit-ups, presses to handstand, pirouettes, flips, splits, and holds. Bike, run, swim, row, etc., hard and fast. Five or six days per week mix these elements in as many combinations and patterns as creativity will allow. Routine is the enemy. Keep workouts short and intense. Regularly learn and play new sports.'

GREG GLASSMAN (FOUNDER OF CROSSFIT™)

CrossFit™ abbreviations and common vocabulary

If you do fancy venturing into the world of CrossFit™, then it helps to equip yourself with the some of the language that's often used (all of which is also available on the CrossFit™ website).

AMRAP – As Many Reps/Rounds As Possible

BP – Bench Press

BS – Back Squat

BW – Body Weight

CLN – Clean

C&J – Clean and Jerk

DL – Deadlift

DUs – Double-unders (When using a jump rope, the rope passes under your feet twice between each jump.)

EMOM – Every Minute On the Minute (So for instance, on the WOD whiteboard one day is written: '10 push-ups EMOM for 10 minutes'. This means that you must do 10 push-ups at the top or beginning of every minute for 10 minutes.)

FireBreather – An elite-level CrossFit™ athlete.

FS – Front Squat

Girls – Several classic CrossFit™ benchmark workouts are given female names.

ME – Maximum Effort

MP – Military Press

MU – Muscle-ups (While hanging from rings you do a combination pull-up and dip so you end in an upright support.)

OHS – Overhead Squat (Full-depth squat performed while arms are locked out in a wide-grip press position above (and usually behind) the head.)

PC – Power Clean (barbell)

PP – Push Press (barbell)

PR – Personal Record

PSN – Power Snatch (barbell)

PU – Pull-ups, possibly push-ups depending on the context

Rep – Repetition (One performance of an exercise.)

RM – Repetition Maximum (Your 1RM is your maximum lift for one rep. Your 10 RM is the most you can lift 10 times.)

Set – A number of repetitions

SN – Snatch

SQ – Squat

T2B – Toes to pull-up bar

TGU – Turkish get-up

WOD – Workout of the Day

P90X™

That it involves 90 days of hard work is all you really need to know about P90X™, the fantastic system that has captured the imagination of millions of people around the world. The star of the regime is Tony Horton, the master of motivation behind P90X™ – the number-one fitness infomercial in America – whose on-screen presence is the perfect mixture of encouragement, humour, discipline and fun. Guys who just shout at people usually have nothing important to say. Tony, by contrast, has a more gentle aura, the kind of power that can stimulate action with nothing more than a glance.

So what is P90X™? Well, it is HIIT, but their USP is what they call 'Muscle Confusion', which they explain by saying: 'By providing an extensive variety of different moves that take time to master, P90X™ is continually challenging the body's muscles into new growth. The more you confuse the muscle, the harder your body has to work to keep up; the more variety you put into your workout, the better and faster your results will be.'

In basic terms, this is very similar to a method that PTs have been using for some time called PHA (Peripheral Heart Action) but with a little Hollywood sparkle thrown in. The PHA system requires us to design an exercise routine whereby each exercise alternates between focusing predominantly on the upper or lower body, the goal being to make the circulatory system constantly having to work hard to get the required blood/nutrients/oxygen to where they are needed. It is very effective because: 1) the fatigue you generate from one exercise has gone by the time you need to use each major set of muscles the next time around; and 2) while being a very intense session, it feels achievable because you can keep telling yourself that you have almost finished working those exhausted muscles.

For either of these methods, there is very little reference to how many reps you should be doing or optimal speeds. In fact, if detail is your thing then you won't like P90X™; the goal is far more AMRAP (As Many Reps As Possible) rather than being focused upon technique and form. It expresses an attitude that I have, which is: 'Do no harm, but rather than stand around discussing what's optimal just get a move on because if you're moving you're improving.' If you buy P90X™ you'll receive 12 DVDs, which are designed to be followed as many times a week as you can manage. The entire regime revolves around old-school bodyweight and dumbbell moves. There are also some yoga and t'ai chi-inspired sections to be performed at different points in the 90-day schedule (these are provided to help prevent over-use injuries, because 90 days of working out at this kind of intensity is tough, and beginners need to be given some breathers to ensure that the connective tissues such as ligaments and tendons have enough time to recover/develop).

The Little Method

This programme is named after researcher Jonathan P. Little – whose specialist field of study is exercise and the prevention/cure of diabetes – and is somewhat catchier than its original full title: 'A practical model of low-volume high-intensity interval training induces mitochondrial biogenesis in human skeletal muscle: potential mechanisms'. The Little Method it is then.

Published in 2010, the study by Dr Little and his colleagues aimed to measure the effects of HIIT when applied in a 'more practical model' (i.e. than that outlined by programmes such as Tabata™ Protocol, which, if done by the book, is a brutal workout). These guys are serious scientists so they would have been looking for optimal results/maximum adaptations rather than the development of a sexy commercial product, and the aim of the study was simply to establish if there is a way of achieving results at lower intensities with less volume than the frequently used Wingate tests (these are lab tests that involve a person performing maximal power-output sprints on an ergometer bike).

Considering how often the Little Method is banded around it was actually a very small trial, but that's not to say the results aren't credible. It involved seven men of approximately 21 years of age who performed six training sessions over two weeks. Each session consisted of 8–12 × 60-second intervals at 100 per cent of peak power output elicited during a ramp separated by 75 seconds of recovery. (A ramp session is a test or workout where rather than staying a constant intensity the challenge is continuously increased or ramped up – on a treadmill this would involve increasing both the speed and incline, while on an exercise bike the rider would increase their speed and also add more resistance).

Prior to the experiment, all the men had been active but were not competitive athletes (Dr Little's goal was to create a system that could be replicated by ordinary people not just conditioned athletes). All of the participants showed an improvement after just two weeks, with their muscle metabolic capacity increasing, which in effect means that their bodies were functioning and processing fuel more effectively. In addition, their functional performance was demonstrated to have improved over the test period simply by the fact that they found the sessions easier to complete.

The testing procedures were done at a cellular level by means of pre- and post-regime biopsies, these demonstrated that after the two-week training regime all the participants were processing protein and glycogen more effectively. If this had been commercially motivated no doubt the results would have been extrapolated to demonstrate reductions in the participants' body fat levels, however there were no claims made in this area. That said, you can safely assume that if the participants continued at the same level of activity for more than two weeks they would have noticed an increase in lean muscle and a reduction in surface body fat.

These results demonstrated that a low-volume HIIT programme can be effective for improving muscle metabolic capacity and functional performance, and in particular that rationing of recovery isn't always a prerequisite of successful HIIT sessions.

The Little Method is a low-volume high-intensity interval training system that induces mitochondrial biogenesis (the process where cells increase in size and number) in human skeletal muscle and it works like this:

Perform 60 seconds of high-intensity exercise (any of the exercises listed in this book working at around 100 per cent of your capacity), followed by 75 seconds of absolute recovery, which means you can sit down, lie down or just keep ticking over.
Repeat 8–12 times.

This means that (excluding your warm-up time) the whole thing will be over in 18–27 minutes. This is much longer than the time spent doing the original Tabata™ Protocol method, but without doubt it's more do-able by people who wouldn't class themselves as being athletic.

Although this study was very small (only seven participants were involved), the researchers found that this approach to HIIT improved the capacity of the participants as measured during time trials and it is now a commonly used training method.

Les Mills Grit™

The fitness dynasty that is the Mills family from New Zealand has an answer for every fitness goal. Want to get stronger? Try Body Pump. Want to improve your core and well-being? Try Body Balance. And so the list goes on, through Body Attack, Body Combat and BodyStep – you name it, they have a workout with the word 'Body' in front of it.

Les Mills is, however, probably a name more familiar to group-exercise instructors than the public because their business is the most successful when it comes to the production of what we call in the fitness industry 'pre-choreographed workouts'. Their package offers a tried-and-tested formula of easy-to-follow pre-formulated sessions that instructors learn by heart, with the music to match the moves and an in-club marketing package for the health-club operator to use to promote the sessions. Access to these materials is via a licence fee (the club pays a fee to use the class content), and the instructors pay a fee to the club to receive and use the choreography and music. This is updated every three months to keep things fresh. So, bearing in mind that Les Mills has either created or been quick to follow every big fitness trend since the start of the 1990s, it's no surprise they have their version of HIIT. This time, though, there's no mention of the word 'Body' in the title. Instead, it's a different four-letter word: 'Grit'.

These Grit sessions are about 30 minutes long and are broken in to five or six tracks (a track is a piece of music that's mixed with sound effects and a strong beat to set the correct pace), as outlined below:

Track 1: accelerated warm-up to prepare for training;
Track 2: propulsion training during which the heart rate begins to spike to its maximum;
Track 3: speed training designed to activate fast-twitch muscle fibres;
Track 4: maximum effort featuring anaerobic exercise with aerobic phases;
Track 5: the core track with an integrated cool-down.

In classic Les Mills' fashion, in order to ensure ownership of their concepts/methods the actual intervals don't use any recognisable formats, so they neither follow the Tabata™ 20/10 seconds, nor the Little Method of 60/75 seconds, etc. Some of the intervals are 1 minute followed by a minute's recovery, while others can be much longer at 3 minutes. Not that this matters, since a Les Mills session is always part science and part exertainment, with the music and attitude of the instructor being a key part of the experience.

Turbulence Training™

Turbulence Training is all about speed. In fact, it's so fast there isn't even time to say the entire name, so it always gets referred to as 'TT'. Craig Ballantyne is the man behind the programme and his magic number is 90 – standing for the 90 minutes' exercise a week you sign up for when you take on TT. An American PT trainer, Craig doesn't hold back with his views about the rest of the fitness industry – long, slow cardio is evil and the like – and in the TT videos he looks straight down the lens of the camera and, unlike the mesmerising Tony Horton, he basically tells viewers to go hard or go home (although if you are doing TT then you are probably already at home because the entire programme is sold via the web as a home-workout solution).

In reality, TT doesn't differ all that much from most of the other HIIT programmes available online. The raw ingredients are the same: weightlifting and bodyweight moves making up the lion's share of the workout, buffered with cardio components. The timing intervals aren't remarkably different either, although they are much longer than the likes of Tabata™. TT goes for 8 heavy reps of weightlifting sets followed by 1–2 minutes of cardio, the sessions are a maximum of 45 minutes and the recommended frequency is three times per week. Personally, I think alternate days would work well, but obviously that means the training days change each week.

So if TT isn't much different from many of the other HIIT programmes available online, why is it so successful? Answer: it's brilliantly presented. Note I say 'presented', not 'packaged'. This is because everything revolves around the testimonials of people for whom TT has worked (with the obligatory shirt-off before and after photos) as well as countless statements relating to research studies by very respected scientists and universities (it's worth noting that most of these statements refer to workouts 'like' TT or present themselves as if the quote were made specifically about TT, whereas this may not be strictly the case. All this means is that TT has been extensively tried rather than specifically tested. I personally don't have a problem with that and think if you can motivate yourself to stick to the programme and be prudent with your nutrition then the chances are you'll also end up on the TT website with your shirt off saying 'this could be you'.

Sprint 8®

It's confession time on this one. Dr Mercola and Phil Campbell are very familiar faces to me since their programme, Sprint 8® features on the equipment manufactured by Matrix Fitness, one of the global companies to whom I deliver master-trainer and brand-ambassador services. However, all of the Sprint 8® research and protocols were developed before that relationship began and they received no specific grant from any funding agency in the public, commercial, or not-for-profit sectors to perform the study, so my views on Sprint 8® are based on their merits alone rather than any commercial bias.

Sprint 8® is probably the most researched and substantiated exercise system that you've never heard of, and I think it is the most universally suitable for 'real' people who have limited experience of exercise or are in fact absolute beginners. Phil Campbell is a passionate advocate of using the protocol, which was developed in three medical centres in the USA: King's Daughters Medical Center, Brookhaven; University of Mississippi Medical Center, Jackson; and Copiah-Lincoln Community College, Wesson.

The Sprint 8® method has evolved from the concept packaged in book form as Ready, Set, Go! Synergy Fitness for Time-Crunched Adults by Phil Campbell, which was written more than a decade ago. It was shown to fight obesity in both an economical and time-efficient manner by naturally stimulating significant growth hormone (GH) release. GH-serum levels are known to increase substantially after exercise, which is beneficial since the hormone initiates lipolysis (fat breakdown), inhibits the uptake and storage of other lipids, and induces muscle hypertrophy (growth).

The research consisted of an eight-week study of free-living individuals (that's geek-speak for normal people), who completed eight hours of exercise over an eight-week trial (20 minutes, three days per week), without dieting. The results were awesome: among the 22 participants, growth hormone values increased 603 per cent following the initial bout, and by 426 per cent by the final bout. On completion of the trial, reductions of body fat in some cases had dropped by as much as 27.4 per cent, with no dieting! LDL cholesterol (the undesirable stuff) also dropped, while HDLs (the desirable version of cholesterol) increased by 2.0 per cent.

Sounds great. So how can you do Sprint 8®? Easy, you can use any cardio equipment (bike, rower, cross trainer, running machine, etc.). If you have access to Matrix Fitness equipment, it may have Sprint 8® pre-programmed as an option on the menu, which means that you simply select the programme and then you get a 'hands-free' workout as the machine will make all the adjustments to incline and resistance for you. All you have to do is go fast. If you don't have Matrix kit then you'll have to press the buttons more often, but the long and short of it is you are going to work as hard as you can for 30 seconds and then cruise (recover) for 90 seconds and repeat this eight times. How hard? As hard as you can! The probability is that if you are programming the CV kit yourself then you will most likely go too fast too soon and struggle to complete all eight repetitions. If this happens, don't think of it as a failure – it just means that you are simply learning how your body functions.

Sprint 8® stands out as a medically developed programme. It can be used by beginners all the way through to elite athletes, and it can be done using CV equipment, bodyweight moves or strength equipment such as dumbbells and barbells (in fact EVERYTHING that I advocate in the workouts in this book).

Here's the summary:

• At least 3 minutes of warming up;
• 30 seconds of flat-out sprinting (Flat out means that if it you were told to work at that speed for 35 seconds you couldn't do it);
• 90 seconds of active recovery.

Repeat this eight times and then rest in the knowledge that your HGH is spiked for at least the next two hours and is chasing down your body fat like a heat-seeking missile.

"Gains include effective weight loss, improved cardiovascular health and increased endurance levels ... "

SoulCycle™

Founded in 2006, SoulCycle™ is without doubt one of the most financially successful yet disruptive boutique workout models on the planet. In fact, SoulCycle™ is disliked by many group-exercise cycle instructors, but I doubt the disdain of some quarters of the fitness industry is going to cause them to lose too much sleep, especially since their A-list riders include celebrities and social-media heavyweights such as the Beckhams, the Clintons, Lady Gaga, Oprah Winfrey and a loyal, vocal fan-base of some the coolest bloggers on the planet.

So what is it and, more importantly, is it HIIT? Well, the clue is in the name: it's all about Soul. Instead of focusing upon cycling technique, form and the production of power (watts), they (in their words): 'Believe that fitness can be joyful. We climb, we jog, we sprint, we dance, we set our intention, and we break through boundaries. The best part? We do it together, as a community.'

Sound cool? Well, it gets better because the instructors range from hippy yoga chicks to snarling, tattooed muscle men who push and pull you through a ride that makes you sweat, smile and maybe even cry for mercy or joy. Oh and did I mention the studio? Well, let's just say every time I've been there I come out smelling of grapefruit because they love to pump its essence into the air while you ride. In other words, the studios are gorgeous, and the music is equally awesome.

But is it HIIT? It should be, but I expect for many of the riders it isn't, simply because they don't put enough resistance on the flywheel of the bike to propagate the physical responses within the body that spike HGH. That's down to rider error, though, rather than the class itself, and since the format of these varies depending upon which instructor you get on the day, they without doubt provide a really fun way to work out.

Orangetheory Fitness™

My first experience of Orangetheory (OT) occurred in California at a session in which the instructor not only wore head-to-toe orange get-up, but also had matching dyed orange hair with the OT logo shaved in. Based on this evidence, I think it's safe to say that orange is a key part of the corporate fabric of the amazing franchise …

What makes this method unlike the majority of boutique HIIT facilities (apart from the preponderance of the colour orange) is that OT holds itself accountable during every session it ever delivers by means of chest-worn heart-rate monitors. These transmit data back to TV screens mounted high up on the walls of the studio to display not just the heart rate of all the class' participants but also to translate the different participants' live pulse into colour-coded zones and calculate the number of calories burned.

The Orange 60 sessions last an hour and roll through bursts of activity and recovery using running machines, rowers and the TRX suspension trainer. During the 60-minute workout, you perform multiple intervals designed to hopefully produce 12–20 minutes of training at at least 84 per cent of your maximum heart rate, or in the 'Orange Zone' or 'Red Zone', in Orangetheory parlance. This is one of the five zones that lights up next to your name on the screen, allowing you to stay tuned in to intensity but not become obsessed with it. Each of the zones represents the following:

Zone 1 is 50–60 per cent maximum heart rate and makes you glow a miserable grey colour;
Zone 2 is 61–70 per cent maximum heart rate, which is a rather cool/tepid blue;
Zone 3 is 71–83 per cent maximum heart rate, and turns your tile on the screen green (green is good, but orange is awesome);
Zone 4 is 84–91 per cent maximum heart rate and turns your tile orange;
Zone 5 is all-out effort of 92–100 per cent maximum heart rate and tips you into the red.

Barry's Bootcamp

According to Barry Jay, his Barry's Bootcamp venues don't just do workouts, they do the ... Best Workouts in the World® and they have been doing so since 1998, long before HIIT became popular.

The signature workout at a Barry's Bootcamp is a super-loud hour-long workout in a dark space that is filled with running machines and steps/benches alongside rubber bands, medicine balls and dumbbells. Every workout is different but should include 25–30 minutes of interval cardiovascular running-machine routines and 25–30 minutes of strength training. Instructors, muscle groups and even workout segments vary throughout the week so that no one class is ever the same. What I like about Barry's Bootcamp is that while there isn't a great deal of validation behind the session plan, if the instructor is even vaguely good you should get a great HIIT workout. Each day on the timetable is aimed at invoking a HIIT effect via the running-machine running sections as well as targeted sections for body parts that change daily. For instance, in a week, you may work on the following:

Monday: arms and abs
Tuesday: butt and legs
Wednesday: chest, back and abs
Thursday: hardcore abs (I guess abs features highly on the clients' 'wish lists'!)
Friday/Saturday/Sunday: everything they can throw at you, since it's a full-body workout in every session.

Barry claims that:
'Our innovative technique works to "shock" the body in the most efficient and effective way to improve your cardiovascular system, lose weight and build muscle. Our world-class instructors are the best in the business and promote a positive, supportive environment that will help you break past your own personal boundaries. Come discover why Barry's Bootcamp has been voted "The Best Celebrity Workout" by Allure, Los Angeles Magazine *and many others. Regardless of skill level, you can burn 1,000 calories in just one hour. You will see and feel results right away in a thumping music-filled environment where every class feels new, fun and exciting.'*

LEARN IT, THEN WORK IT!

For a number of years now the two biggest fitness-industry organisations – International Health, Racquet and Sportsclub Association (IHRSA) and International Dance-Exercise Association (IDEA) – have both placed HIIT on their prediction lists for the next major fitness trends. It is worth noting, though, that the trends that push through the noise and become mainstream invariably generate some negativity, either from people who don't like change or policy-makers who have to mitigate and manage people's expectations (or in other words ensure that common sense prevails). These voices of reason include such associations as the American Council on Exercise (ACE), American College of Sports Medicine (ACSM) and National Exercise Trainer Association (NETA), and these are starting to raise concerns that HIIT is being seen and marketed by many as being the only form of exercise worth doing (inflexible mantras such as 'go hard or go home' spring to mind) and could therefore put off participants who would benefit from a lighter form of activity.

Their remit, unlike that of some unscrupulous fitness providers, is to safeguard the long-term welfare of the participants. For instance, ACE certifications employ something called an Integrated Fitness Training (IFT) model for functional movement, which describes how fitness professionals should carefully and systematically progress people from functionality, to health, to fitness, to performance. This is a critical concept with respect to high-intensity protocols, which tend to quickly push for full-blown performance. So, although HIIT works, it is imperative that a fitness professional ensures the participant is pre-conditioned and already has the stability, stamina and strength in their physical make-up before they start doing performance protocols (HIIT).

Richard Cotton, the national director of certification for the ACSM, concludes that it comes down to conscientious, qualified instruction; comprehensive participant screening; and proper programme development and modifications. 'Any exercise

program involves risk,' he says, 'but it can be mitigated with adherence to policies and procedures that are consistent with accepted standards of practice.'

This means that fitness instructors must remember that unfortunately many people walking into fitness facilities are de-conditioned, and so need to be screened and then directed towards the most appropriate methods of exercise for their health profile – which may not necessarily be HIIT. The reason for this, explains Michael Iserman, the director of personal training for NETA, is that: 'The incidence of negative outcomes increases during high-intensity exercise, and injuries and adverse cardiovascular events occur, even during thoughtful, well-designed programs.'

In other words, bad stuff happens sometimes, and exercising exposes everyone to some risk of injury, especially if it is done at high intensity. Unfit people may get injuries as a result of previous inactivity and super-fit people get injured especially as they increase their exposure to new challenges – even Olympic athletes with the finest coaches in the world suffer injuries, despite having the best support structure available to them.

Careful consideration and some caution, then, can only be a good thing, but there is a fine line between doing something and doing nothing – every PT client that I have ever worked with has had 'different buttons' that needed to be pressed to get 'results'. There were some who were always looking for a reason to give up (DOMS would have been the end of the world for them), while others would have considered asking for a refund if they were still standing at the end of a session. As a reader, though, without a PT standing next to you, you have to make your own judgement call on your level of fitness and how hard you should push yourself.

Once you have done this self-assessment, it's time to get HIITing, which is where my second mantra (the first is 'if you're moving you're improving') comes in: 'Learn it, then work it'. This means that rather than going full-steam ahead in the first workout session, you should instead practise drills and movement patterns. In short, you should be working at perfecting technique and developing a baseline level of fitness before adding intensity. This will prevent injury and ultimately optimise results.

All of this rhetoric is aimed at persuading you to ensure that you take the time to learn the exercises and treat the movements as an individual skill, before you go at them as part of a complete workout. Improve slowly and one rep at a time before you attempt them flat out or try to do As Many Reps As Possible (AMRAP).

PART TWO

WHAT YOU NEED

HIIT KIT

HIIT clothing

Your clothing during a HIIT session is important, because wearing the wrong thing could affect your enjoyment and prevent you from gaining optimal results.

This book is focused on workout gains and, in light of this, fitness fashion is a minor consideration (although some members of the fitness fraternity might not agree). What people wear to work out is of course up to them, and ranges from those wearing 'badge of honour' shirts that they have donned for every session for the last 10 years, to those at the other end of the spectrum – the 'sports luxe' (sports luxury) brigade who wear different fashion fitness outfits at each gym outing, since for them image is seemingly everything. Everyone is different, then, but no matter what it looks like, if you follow the guidelines below you will be all set for a successful and, most importantly, safe HIIT session.

FOOTWEAR

Starting from the ground up – if you get the shoes right, everything else will follow. Ensure that your trainers/sneakers are fitness-specific (Reebok Nanos are my weapon of choice) rather than being specifically designed for running, since the latter type often have too-tall sides for certain types of HIIT move.

There are people who advocate being barefoot or wearing minimalist shoes, and these are also fine. However, if you are new to this style of shoe or lack of strength in your feet, give yourself time to adjust to decreased support and cushioning by wearing them or going barefoot alternately with your trainers or sneakers.

LOWER-BODY CLOTHING

Shorts – cotton, Lycra or otherwise – let you move more freely, and since many HIIT moves require you to get up and down very quickly, unimpeded movement around the knees is essential. Another consideration is body warmth, and while I personally prefer shorts to long track pants or full-length leggings for exercising, it is important to ensure you don't get cold before or after your workout, so you may want to consider some warm, dry (your shorts may well get damp with sweat!) trousers of some kind too.

UPPER-BODY CLOTHING

The upper body options when doing HIIT are a T-shirt or tank top, preferably close fitting and made from wicking material. Again, body warmth can be a consideration so bear in mind your workout conditions and wear layers if the conditions are on the cool side. Personally, I find that you can happily forget that you are wearing a tank top, whereas with a T-shirt you often find that they either ride up or that you are constantly having to adjust the sleeves or shoulders. Such distractions are unhelpful when giving it your all in a HIIT session.

SPORTS BRAS

Ladies, you will be moving very fast at times and also changing direction quickly and that means, depending upon the size of your breasts, that you need to ensure you are wearing suitable chest support to be ready for action. Making the right choice will not only inspire you with confidence but it will also look after the very important Cooper ligaments that support the female breast. These ligaments can be stretched if you do high-impact activities without the appropriate support, and over time they won't snap back, resulting in sagging breasts. So, even if you aren't interested by fashion, think about your Coopers.

There are three possible bra options:

1 If you have relatively large breasts, a highly supportive sports-specific bra is essential. These are usually engineered with metal/plastic wires and will provide necessary support and control during movements. It is advisable to get this properly fitted by a specialist.
2 If you have medium-sized breasts, you should go for a tight sports bra that is an adaptation of the boob tube. This is like a thicker, more substantial version of a bikini top and effectively flattens breasts to keep them closer to the ribs and therefore controls how much they move.
3 If you have small breasts, you can go for a racer-backed bra top. These are similar to the style used by Olympic beach-volleyball teams and will keep everything firmly in place.

HIIT tech

We all love gadgets and these can play an important role in measuring and evaluating our exercise output. Popular options include exercise monitors worn on the wrist, waistband and even around the neck as a pendant. As fun as these devices are, though, it is important to question the accuracy and the relevance of the information that they collect.

One gadget I rate highly is the MYZONE® chest strap. This monitor tracks your workout and effort and lets you study your performance through its smartphone App. The chest strap you wear while working out is light and discreet and won't get in the way, unlike watches, which can be damaged or destroyed during workouts when they have come into contact with kettlebells or the floor during HIIT moves such as squat thrusts.

The MYZONE® claims to be 99.4 per cent accurate, which is as accurate as the equipment used in hospitals to monitor heart rates. Moreover, unlike many of the big-name devices that measure your heart rate by taking the reading at the wrist, the MYZONE® chest strap, due to its proximity to your heart, measures every beat. Wrist devices are notorious for losing the odd beat and therefore are generally less accurate. Apps such as the one that accompanies the MYZONE® chest strap can greatly enhance the workout experience and you'd be amazed at how fun it can be to chart your progress over a number of workout sessions.

The downside to not wearing a watch during HIIT sessions, however, is that you need an alternative method of timing how long you are working and recovering for. In the gym, this is unlikely to be an issue since you can look at the studio clock, but when you are working out at home or outside it is best to opt for an App to do the timing for you. There are hundreds to choose from on the App store for both Apple and Android users.

A personal favourite is a brilliantly simple App called Tabata Timer. It stands out because when you play music through the same device, it not only times the workout and recovery periods but it also throws out a voiceover above the music to give countdowns and warnings that it's time to go and stop. Another good one comes from Escape Fitness called TIME2TRAIN and is available at any App store.

HIIT gym-based equipment

The following gives some background to the kit that will be used for the gym-based HIIT workout programmes. It is important to get to know the equipment in advance, and this includes knowing your own unique set-up for the machines – 'learn it, then work it'.

The variety of kit found in gyms around the world is extremely eclectic. Your local one may have a huge selection, but most of it serves no purpose if a HIIT session is your goal. An easy way to work out what you do and don't need is to first dismiss all of the weights machines that you sit or lie on, then ignore anything that is predominantly intended to develop balance and coordination.

This book only features equipment that every gym should have. Moreover, in addition to being 'common', these have been selected because they also facilitate a wide range of exercises and challenges that are perfect for HIIT. The key pieces of equipment are: a TRX; a rig; an Olympic bar plus discs; and a plyo box.

TRX

The TRX is also known as a 'suspension trainer'. It's basically an adjustable set of straps that you use to make your own bodyweight challenge your muscles while you move through different exercises. There are plenty of alternative brands of suspension trainer out there, but the TRX is arguably the best on account of the fact that the straps offer the perfect blend of quality, functionality and safety.

There are hundreds of exercises that can be performed using a TRX, but this book includes only those that achieve maximum effects in the minimum of time and, most importantly, require only a small amount of skill.

Here are a few guidelines to follow when using a TRX:

• No slack on the straps – if the straps lose their tension then you aren't suspended and that means you are basically standing still holding a couple of handles for no reason.
• When you are standing up, moving your feet closer to the anchor makes the exercise harder (the 'anchor' is where the straps connect to the wall or frame).

RIG

Rig or 'functional' frames are increasingly present in gyms. A rig could range from being a very basic 2 x 3m (6ft 5in x 10ft) metal frame that resembles scaffolding poles, all the way up to the enormous frame called The Queenax, which is taller, wider and probably more expensive than any other rig available. The best versions enable you to do hundreds of different bodyweight exercises without the use of additional equipment.

Here are a few guidelines to follow when using a rig:

• If you are new to rig training, first work on your grip. You can do this by standing under the bar and hanging with straight arms.
• Once you are able to hang for 10 seconds, try to add some 'pull' to the action. If you keep working at it you'll be doing full pull-ups in four to six weeks.

However, to benefit from using a rig quickly, I've developed a fast-track method that incorporates a jump from the floor, so that rather than isolating the biceps and making them do all the hard work of pulling your chest up to the height of the bar, your legs provide some thrust. This is because muscle isolation has no place in a HIIT workout.

OLYMPIC BARBELL AND WEIGHT DISCS

There are barbells and then there are Olympic barbells. Regular ones are short, solid bars that have either fixed weights on each end or small removable discs held in place by collars.

Olympic bars are much longer – 2.2m (7ft 2in). The bar itself is assembled from multiple components, the most important of which are the ball bearings that are mounted inside the thicker section at the ends of the bar. These enable the lifter to flick their wrists and make the bar spin, which means that you can move/lift the bar very quickly because the weight plates/discs don't rotate but the bar does. Without this feature, not only would you have to lift the weight up but you would also need to generate torque to rotate the weight plates, which, trust me, is impossible to do once the weights creep above 30–40kg (66–88lb).

As the name suggests, this style of bar is used in the Olympic Games for competitive lifting so they are designed to cope with massive amounts of load (always check the marking on individual bars for the maximum weight limit). However, for the purposes of HIIT we'll be using the Olympic bar with a fraction of what you could potentially lift if you were doing a competition-style One Rep Max (ORM). I use as little as 30 per cent of my ORM for some HIIT moves and never more the 60 per cent, because otherwise you just grind to a halt.

Here are a few guidelines to follow when using an Olympic bar:

• Always make sure that you have room in front, behind and ABOVE you before you start lifting an Olympic bar.
• Don't lift a bar without fitting a set of collars to hold the plates in place. It's lazy not to and you'll either hurt yourself or send the weights crashing off the end on to an unsuspecting gym user.

PLYO BOX

Plyo or 'plyometric' boxes come in many shapes and sizes and are made of a multitude of materials. Personally I prefer to use the soft (foam) ones as they are more forgiving than other materials and are less likely to damage shins. If you are using solid boxes, start with a low height and work up the bigger options once you have perfected your technique.

Plyo boxes are one of those products that have become increasingly present in mainstream gyms. They originated behind the Iron Curtain when Russian and East German gymnasts started to introduce a 'rebound jump' into their training, a move that involved them jumping off high objects so that they could use the impact and inertia they experienced upon landing to strengthen and condition muscles ligaments and tendons ('rebound jumping' because the goal is to rebound off the floor as soon as you make contact with it).

As a guide, any plyo box that is higher than your knee joint is high enough to be productive, but anything taller than hip height is great for one-rep jumps but not so good where the intention is to complete multiple reps one after another at speed. As ever, work on technique before adding intensity, so start low then get higher once you can almost do the work in autopilot mode: 'Learn it, then work it.'

Here are a few guidelines to follow when using a plyo box:
• Make sure there is enough space above the box for you to be able to stand up.
• Also ensure that if the manoeuvre goes wrong there is space behind you to safely land, even if it means landing on your butt.

RUNNING MACHINE

There are many new innovations concerning running machines or treadmills and although this is fantastic, this book focuses on moves that can be done on conventional ones to ensure that you'll have access to the tools needed to perform those exercises.

You'll also notice that I don't use the term 'treadmill', which sounds so negative. Instead, I prefer 'running machine'. My advice when using one is that you don't watch TV and don't look around you. Look ahead and keep up. Think of yourself as constantly being in the lead rather than simply keeping up with the running belt; that way you'll run with style, grace and, dare I say it, elegance.

Using a running machine

When doing interval training on a running machine you will be blending together fast and slow, and flat and incline settings. It can be a common sight in gyms to see people doing fast sections of a run then, when it comes to their recovery period, grabbing the crossbar of the running machine and jumping their feet on to the side rails, standing there and recovering until the next interval, at which point they jump on to the moving running belt. Although to some people's minds this technique may be considered the 'normal' way of doing intervals, it can't be condoned since it isn't very safe.

Best practice is to use the speed controls then, when you need to stop running for a recovery interval, press the PAUSE button on the control panel. This will slow down the machine to a stop but leave the timer/clock running. Once your recovery period is almost up, deactivate the pause feature, start walking and speed up the machine until it's moving at your desired pace.

If you treat running machines with common sense and respect then you should be fine.

SPIN BIKE

Spin class, spinning session, ride session, studio bike, power/watt bike ... These are all terminologies that get used when talking about training sessions that happen on a bike fitted with a weighted flywheel and a friction/magnetic or air-resistance system that allows the rider to pedal against variable amounts of load.

'Spinning out' is a phrase that cyclists have used for decades to refer to a style of riding that involves them spinning the wheels/pedals at 90-plus RPM against a light resistance, but the word 'Spinning®' is now a registered trade mark. This style of class has experienced a massive surge of popularity recently and has been most successfully packaged by US-based Soul Cycle™.

For the purpose of HIIT, I like to use spin bikes in a stripped-back fashion and, most importantly, always ride against plenty of resistance. The way the bikes are designed means it is possible to ride them very fast with hardly any energy expenditure – and many see leg speed as a status symbol – but getting the RPM (revs per minute) above 120 only really develops skill rather than power and is not much use for HIIT. Moreover, being able to spin fast is a skill that has no immediate benefits to the rider when they are off the bike, and is therefore to my mind rather futile.

That said, I do on occasion recommend riding at levels of 135RPM, but only if there is a noticeable amount of resistance to push against. This is because basically speed is speed: it's only productive if it's mixed up with some resistance to generate power.

Another important element of a spin session is the position-based moves. These include lifts, 'jumps' and 'hovers', which are used to enhance the workout.

Modern gym bikes (including the one illustrated in Part 3) have a number of important features:

• Many have a console that starts to flash if you exceed 150RPM. This is to warn you that you are placing yourself in a world of unproductive danger.
• Some also have lights on top of the console that glow different colours depending upon how much power you are generating in real 'time' (the light spectrum goes white, blue, green, yellow, red). You'll notice that in all the photographs of spinning in this book, the red light is showing. For the purpose of HIIT, anything less would be a compromise, whereas in a regular endurance session you are likely to be working at approximately 25–30 per cent less intensity/power and would expect to see the green/yellow lights glowing.

Using a spin bike

There are many different ways of adjusting the seat and handlebar height, and the best bikes also have a feature that enables you to move the seat bar both forwards and backwards.

Set the seat height so that your legs never lock out when you turn the pedals. If you have the option to move the seat forwards, slide it towards the handlebars. However, it is important not to adjust it so that your knee passes in front of your toe when the pedals are in the 3 o'clock position. At 12 o'clock, your foot is at its highest point and at 3 o'clock it's in the forward position (which is also known as KOPS or Knee Over Pedal Spindle). At 6 o'clock, your foot is at its lowest point of the pedal circle and, finally, in the 9 o'clock position the foot/leg is at the back of its revolution and probably simply getting back to the top (or 12 o'clock) ready for the next push rather than actually generating any power.

Here are a few guidelines to follow when using a spin bike:

• Check if your bike is 'fixed wheel'. If it is then that means it is always in gear, so when you are riding you can't just stop pedalling like you can on a road bike – the bike pedals will keep turning due to the power that is being generated by the flywheel.
• Until you are a hardened rider, the skinny seats on spin bikes can all feel like torture devices. This is because they aren't really there to be sat upon; they simply give you somewhere to perch when pedalling. As you get used to riding this style of bike, however, you'll put more weight through your legs and less into your crotch. Wearing padded cycling shorts also helps.
• A very good philosophy to have at all times when riding a spin bike is to aspire to make the movements of your body and those of the bike blend into one as much as possible.

ROWING MACHINE

There are many styles of rowing machine, but the Concept 2 rower is one of the most common types found in gyms and is perfect for the HIIT workouts found in this book. The mechanisms of this model (and most others) are usually fairly similar, and include:

• A seat that slides on a rail
• An oar/handle connected to a chain that propels a flywheel/damper
• A spiral damper.

The damper is the lever on the side of the flywheel that controls how easily air can flow in and out of the cage/fan blades. This is appealing for HIIT since it enables you to flash between light/fast and heavy/slow. The lever sits next to a scale of 1–10, indicating how much air is drawn into the cage on each stroke.

Higher damper settings (7–10) allow more air into the flywheel housing. The more air that flows, the greater the challenge to keep the flywheel spinning. More air also makes the flywheel slow down quicker between strokes and the knock-on effect of this is that every stroke is a challenge since there is no opportunity to benefit from the power you generated with the previous pull.

Damper setting number 6 is suitable for the HIIT sessions in this book. You should aim for 25–35 strokes per minute. Any more or any less and you can't be sure the technique I outline will work as effectively.

Using a rowing machine

Rowing machine technique is broken down into four phases:

1. The catch: Your arms should be straight and your head should be facing towards the console. Shoulders are low rather than hunched. You are leaning forwards from the hips (hips hinged). Your shoulders, if your mobility/flexibility allows it, should be in front of your hips. Your shins are vertical and shouldn't ever go past this angle.

2. The drive: Initiate the push through your legs. Just before the legs are fully lengthened, lean backwards from the hips and pull with your arms so that your elbows bend and the oar draws level with your bottom ribs.

3. The finish: Your upper body is still leaning back, your shoulders are low/depressed and your wrists are flat. There's no force being generated during the finish phase; it's basically the calm before the storm of the next stroke.

4. The recovery: Extend the arms fully and simultaneously lean forwards at the hips towards the flywheel. Once your hands have cleared your knees, pull against the toe straps so that your entire body slides along the rail of the rower to get you back into position ready for the next Catch.

Perfecting a rowing technique takes practice and at the start it can feel like you are being worked by the rower, rather than you being in charge. Of all the cardio kit in a gym, this machine takes the longest to master, so if you are inexperienced please make sure that you take your time to work on your skills before piling on the intensity. Once again, 'learn it, then work it'.

THE SLEDGE (OR 'PROWLER')

'Prowler' is the slang name for the weighted sledge. They come in various shapes and sizes but the basic format is two smooth runners that slide on the floor with a single horn that you load weight plates on to, then you have two handles that you push against. It's impossible to say how much weight is the right amount to load on to a sledge because the floor surface on which you are pushing it has a massive effect upon how hard it is to get the thing moving and then keep it going.

Using a sledge may be one of the most challenging and effective whole-body exercises in the gym and it's also, despite appearances, very inclusive in terms of the abilities of the people who use it.

The best technique depends upon your goal. If you want to target the entire body, and especially your core, then grip the handlebars as low as you can without falling flat on your face and keep your arms lengthened but with some flex at the elbows. Your shoulders and hips can either be level or you can position your hips slightly lower than your shoulders and upper back.

The best way to get the sledge moving is to use both legs equally for the initial push and then take small steps until you have created some momentum. Now lengthen your strides and keep going until you either run out of track or energy.

You can use the prowler for a brutal leg workout if you load it up until it's as heavy or even heavier than you, then shove it one rep at a time, pushing it equally with both legs then stepping forwards to prepare for the next repetition. However, while this is great for strength, it's not so relevant for HIIT.

There are various things that can go wrong when you use a sledge, and it may not be you who gets hurt – it's often people getting in the way who get injured rather than the user. I've also seen people crash over the top of the prowler. This is most likely due to a lack of confidence as much as lack of skill. It takes self-assurance to hold the handles close to the ground but this is actually better than holding it up high, since this tends to make the front end of the sledge dig into the floor. So, be brave, get your hands low and head down, then PUSH!

HEAVY PUNCH BAG

Health clubs and gyms went through a phase of installing heavy bags (a heavy bag is one that is generally as big as you are as opposed to the much smaller speed bags that you see boxers hitting at hundreds of reps per minute), and they are good bits of kit, if you know how to use them.

The rules of using a punch bag include: no kicking the bag; no punching it with bare hands; and definitely don't try to win. The bag will always be the victor, so let it swing freely rather than leaning against it and chasing it around the room.

Boxers are formidable users of bodyweight exercises so we can learn a lot from them. Aim to keep the feet moving lightly underneath you and make fast jabs at the bag. Also, whatever you do with the right side of your body 'leading', you need to repeat with the left. This feels very odd to some people, but since the goal is rarely to become a proper fighter it's more important to develop balance rather than preserve the natural inclination of being able to punch harder with one hand or the other. There's no need to lean into the bag – if you hit it fast enough you'll very quickly be inducing the type of physical responses that we desire for you to experience during a decent HIIT session.

HIIT fuel and fluids

It seems to me that many personal trainers have morphed into a completely different species from the people they train, both in terms of attitude and eating habits. Their attitude is that it's inconceivable that not everyone has time to prepare 30 meals in advance on a Sunday evening and then store them in plastic boxes ready to be consumed during the week ahead. In reality, life is rarely that simple and most people's eating habits have to work around their commitments, rather than the other way round.

For this reason, it is very useful to have a basic grasp of how to prepare nutritious food as and when required, as well as understanding how and why your body needs certain things, especially when you are training. The following provides some guidelines.

Nutrition to optimise a HIIT programme

To get the most out of any fitness regime we should follow a healthy-eating plan, especially if fat loss or weight control is your goal. It's all too easy to sabotage the benefits of a training session with a post-workout carb-fest of junk food. An effective and well-rounded nutrition programme includes ingredients such as wholegrains, fruits and vegetables, and lean proteins. The best nutrition ones provide carbohydrates to fuel the body and provide energy stores for workouts. It is important at this point to state that I am a vegetarian, and I know from experience that so long as you adhere to a balanced meat-free diet, this will not hold you back.

The supermarket can be a daunting experience if you are trying to be healthy because those store designers have done everything they can to lead you into temptation, so my Golden Rules are there to help you to narrow things down a little when you venture into the store on you next food shop.

NUTRITION EXPLAINED

Knowledge is power, but in the world of sustained health and wellness, insight, consciousness and responsibility are the keys that you need to achieve long-term weight management alongside long-term nutritional stability, so it is well worth familiarising yourself with the basic concepts of nutrition.

Macros (macronutrients)

The word 'macronutrients' is the clinical terminology for the three main sources of energy: protein, carbohydrates and fats. These are essential to health and the three key things that you have to consider if you are going to successfully manage your weight and well-being.

Protein

Most people instantly think muscles when they hear the word protein, but in fact the body needs it for more than just the maintenance or development of a decent set of biceps; protein plays a key role in the maintenance of your immune system and helps manage inflammation of soft tissues. In addition, protein helps regulate the metabolism and influences your BMR (basal metabolic rate) or, in simple terms, the speed at which the body utilises fuel. This is crucial for HIIT. The presence of protein is also responsible for the production of healthy hormones that influence the function of various cells.

Good sources of protein include eggs, milk, fish and seafood, soya, lean meat, poultry and nuts.

Protein powder

The subject of protein can't be discussed in a fitness book without a mention of protein powder. The most common of these is whey powder, which is a waste product left over during the manufacture of cheese, but there are many others. As with everything you put into your body, it is best to get your protein from a source that seems logical and safe, so ensure that any specialist protein product you take has been thoroughly researched.

Carbohydrates

Carbs are the only viable sustainable energy source for muscles during intense exercise, and your central nervous system, kidneys and muscles cannot function without them. And that's not all: 80–90 per cent of your brain's functions are powered by carbs, so no carbs, no brain function. Game over!

In short, getting the right carbs is essential to HIIT success and they are a must-have for every active person. However, we hear about good carbs bad carbs, so are all carbs equal? Well, a carb is a carb as far as the body is concerned, but the reason why there is so much conflicting information is because of all the other stuff that's wrapped up with the carb itself. This means that although carbs that come from white food (flour, rice, potatoes) are pretty much the same in terms of energy content as their more colourful relations, they tend to lack the added goodness of their multicoloured cousins. For instance, brown rice contains more fibre than white, which is good news for digestion; brown or seeded bread is better in so many ways when compared to plain white loaves; and white potatoes can't get near the nutritional heavyweight that is the sweet potato.

Fats

Fat needs rebranding: currently, too many people think of fat as being bad for you. This generalised view assumes, however, that all fats are the same. They aren't. Most of them are good for the body, although there are undesirables that are best avoided.

As with protein, if your diet is devoid of fat then you'll eventually die of malnutrition. Viewed as the body's energy reserve, fat also provides protection (padding) around some of the body's vital organs and nerve routes throughout the body and ensures healthy hair, nails and skin. Moreover, fat is – crucially – a catalyst for the absorption of some essential vitamins (A, D, E and K) and without fat the body can't manufacture some of its essential hormones.

GOOD FATS

Polyunsaturated fats These naturally occurring fats have anti-inflammatory effects upon the body and can be found in oily fish, where they go hand in hand with omega fatty acids (vegetarians can get almost the same hit of goodness from flaxseed and linseed oils).

Monounsaturated fats These are the easy fats for vegetarians and carnivores alike to consume since they are in plentiful supply in foods such as avocado, olive oil, nuts and seeds, all of which in the right volume can help regulate blood-sugar levels and also progressively increase the good cholesterol in the bloodstream.

Saturated fats These are the fats that have been in and out of fashion and are what we think of as the most obvious of all the fats. Once considered to be bad for us, some are now viewed as actually being good, in moderation. They can be found in animal fats, eggs, butter, cheese and coconut oil.

BAD FATS

Hydrogenated fats/trans fats Read the label: if it says hydrogenated fat then I suggest you put it back on the shelf! Beware, since it may be disguised as something that sounds healthy, such as 'hydrogenated palm oil'. The issue with hydrogenated fat is that the body simply doesn't know what to do with it when it's ingested and therefore struggles to process it. To avoid it, don't eat most low-fat ready meals or anything that is marketed as being a tasty low-fat diet product.

Micronutrients (micros)

Micronutrients are extremely important chemical substances that are required in small amounts for normal growth and development, and include calcium, iodine, iron, magnesium, and vitamins A, B, C, D and K. These can be found in a variety of fruits and vegetables, meats and nuts. While in theory you should be able to get all your micros from your food, it does no harm (other than to your purse) to take multi-mineral and vitamin supplements, so just think of it as an insurance policy.

HIIT nutrition 'Golden Rules'

Everyone is unique and we each have different tastes and relationships with food. However if you follow the rules below you should be well on your way to major gains through HIIT:

1 You can eat anything you like so long as it only comes from just three supermarket aisles: dry goods (pasta, rice, tinned fish and pulses, etc.); dairy, seafood, meat and poultry; and fruit and vegetables. This rule will flash-cure the junk-food issue and, moreover, the time and effort it takes to prepare fresh ingredients also will make you have a greater respect for the food that you eat.

2 You can only eat that 'anything' after you have drunk an entire glass of water (500ml (17fl oz)). This will hydrate you and remove your appetite for unhealthy foods. Moreover, many people mistake thirst for hunger, so dealing with the former may stop you from eating unnecessarily.

3 The consumption of alcohol is off limits in the house. This means you can drink socially, but not to pass the time.

4 When you are eating, try to just think about the food and how much you are eating. This means you are forbidden to watch TV, read magazines, browse the internet or talk to people while you are eating. The reason for this is that if you are distracted then you aren't thinking about your food, you probably aren't chewing it effectively, and you will keep putting it into your mouth even if your brain is sending out signals to say that you are full.

5 You have to eat off specific plates and bowls. This is so that you can accurately measure and assess portions to avoid over-eating, something that can only be done if you use the same-sized plates and bowls on a consist basis.

6 Try not to eat later than 7pm in the evening, if at all possible.

It is important to consider when meals should be taken. New research has identified that it is beneficial to leave gaps between meals, and 12 hours between dinner and breakfast, allowing the body to fast for a sufficient period of time between meals. Snacks are to avoided, other than the post-workout recovery snack, and if you find become hungry this can often be alleviated with a glass of water.

Sugar

Sugar, rightly so, is now considered to have no place in a healthy diet. However, to be clear, not all sugars are equally bad. There's the white processed stuff that you might put in your coffee, which isn't great, then there's the naturally occurring sugars in fruits, which at least have many health benefits. However, the type that's currently flashing brightest on the bad-for-you scale is fructose.

Found in huge volumes in most processed foods, there are a number of reasons why you should avoid fructose sugar. First, it's very 'clever' stuff – it 'makes' us want to eat more and more of it, and because humans haven't evolved an 'off switch' for fructose (as it has for other natural foodstuffs) to signal when we've had enough, the brain just keeps saying 'give me more'.

Second, research has shown that it converts to fat in our bodies rather than hanging around in muscles as energy for our immediate needs. Third, it inhibits the immune system, making it harder to fight off viruses and infections and causing you to feel lousy. Worse, as well as being linked with the speeding-up of the ageing process, it has now also been connected with the development of cancer and can cause spikes of activity in the adrenal gland. To top it off, it can also cause hyperactivity, anxiety and a loss of concentration and, in the long term, contribute towards tooth decay and the onset of type II diabetes. It's not great stuff.

That said, despite the fact that it is found in fruits and vegetables, the presence of fructose should not stop you from eating these otherwise beneficial foods, which contain essential micronutrients and fibre. It is also fairly hard to overeat fruit and vegetables and, in general, fruit is a minor source of fructose in the diet compared to the added sugars in processed foods. So keep eating a rainbow of fresh foods.

If you need any more convincing that it would be wise to moderate surplus sugar from non-fresh-food sources completely from your diet (which is hard), then consider that the overconsumption of processed sugar is increasingly being declared by scientists as being as bad for your health as smoking.

Alcohol

Alcohol and HIIT don't really mix, since it causes the body to dehydrate as your system tries to flush it out. As a result, even the smallest amount of alcohol is going to have a detrimental effect upon your 'gains' from doing HIIT, or in fact any type of exercise. Having a hangover is also demotivating, and since you don't sleep properly after drinking, you may well feel lethargic and not put in enough effort even if you do try to do the exercises.

Coffee

Research has highlighted the positive impact that caffeine can have on performance and fat burning. The stimulant has been reported to slow the depletion of glycogen by encouraging the body to use fat as its key fuel, this in turn helps to conserve energy which let's us work for longer. Caffeine has also been shown to increase an athlete's speed, accuracy and levels of endurance. It is important to remember that individuals react differently to varying levels of caffeine, and should therefore be consumed with care and avoided completely within four hours of your planned bedtime.

Hydration

Adequate water must be consumed to ensure complete hydration during workouts. Plain and simple water is best – better than any sports drink – though it can be enhanced with electrolyte tablets. These supplements are dissolved into water and replenish all of the components that sweat drains from your body. After a workout, cow's milk is thought to be the most hydrating drink, and has the added benefit of aiding recovery and helping to build muscles. It may even help to stave off the dreaded DOMs.

Unless they make a conscious effort, most people don't drink enough water. Recommendations on the correct level, and the correct manner, of intake vary (many foods contain a lot of fluid, which counts towards your daily intake), but in general most people feel better if they drink around 2 litres (3 ½ pints) a day. Unless you are cycling in the Tour de France or participating in the Marathon de Sables, you probably won't need that much more than this – up to about 3 litres (5 ¼ pints) is sensible if you've been sweating hard during HIIT. If you take on too much (anything more than about 5 litres (8 ¾ pints) a day), you run the risk of hyponatremia, which is when the bodies salt/sodium levels drop dangerously low.

Assessing your urine is the easiest way to gauge your hydration. If it is anything darker than a pale-straw colour then you need to drink more.

HIIT WORKOUT NUTRITION

HIIT sessions are supposed to be part of your everyday life once, twice, three or maybe even four times per week. Given this range of different levels of energy expenditure, coupled with all the other variables that go into making up a diet that is healthy for you, it'd be futile for me to give you a menu of what to eat and when without turning this into a diet or cookbook. Also, I didn't want to assume that all of you have the goal of losing weight (or dropping fat levels, to be more accurate), which is how most books seem to approach this situation. If that is your goal, then the measures outlined in my Golden Rules (previous page) are still going to be more likely to have a positive effect than me creating a one-solution-fits-all meal-plan solution.

Pre-workout nutrition for HIIT

Due to the intensity of HIIT workouts in general, it's vital to follow a healthy-eating plan with adequate nutrition in the hours leading up to your workout of the day. Try to eat a moderate- to high-carbohydrate meal that also includes protein approximately 3–4 hours before the HIIT workout. After your session, if fat loss is your goal, stay away from the carbs and instead opt for a good-quality source of protein, such as chicken, turkey, fish, eggs or nuts.

Options for a pre-workout nutrition include:

- Whole-wheat toast with peanut butter and banana
- Greek yogurt or cottage cheese with fruit
- A small handful of dried fruit and almonds
- Quinoa and chia seed porridge
- Standard porridge sprinkled with seeds
- Oatcakes topped with cream cheese and banana
- Poached eggs on toast
- A smoothie made with vegetables, banana, nuts, oats and water

Options for post-workout nutrition include:

- Wholegrain cereal with fruit and almond milk
- Whole-wheat crackers with fruit and cheese
- Hummus and pita bread
- Oatcakes topped with cottage cheese and avocado
- Banana, cacao and nut-butter milkshake
- Tinned mackerel or sardines on wholemeal toast
- Hard-boiled egg and a banana

Post-workout nutrition for HIIT

After your session, thoughts should move to replacing energy stores (glycogen) and repairing muscles that have been broken down during the HIIT workout. A widely accepted formula is a combination of carbs and protein (remember though that excess carbs could prevent any fat reduction that you have just triggered via the workout and subsequent EPOC). Dieticians refer to research that uses a 3:1 ratio of carbohydrates to protein, and this should be consumed within 30 minutes of completing a HIIT workout. It might, therefore, be worth preparing something in advance and taking it with you to the gym.

Sleep

Sleep really is incredibly important, so if you possibly can then you should be in bed and ready to go to sleep by 10.30pm every night. Being awake late at night can often lead to mindless consumption of snacks but, more importantly, going to sleep at around 10.30pm is a good way of controlling the stress hormone cortisol, which plays an important role in food cravings and how you process food.

The use of screens such as TVs, smart phones and tablets in the bedroom and around bedtime is to be avoided, because the blue light they emit tells your brain to wake up, doing nothing to help you drop off into quality deep sleep.

PART THREE

THE EXERCISES

BODYWEIGHT HIIT EXERCISES

As you read the exercise descriptions you'll see that I have assumed you fit into the two different types of people I have trained in my career:

Type A who wants to be shown and told what to do and get on with it – you are the guys who get a new phone and just work out how it functions as you go, and will most likely only read THE BRIEF;

OR

Type B who want to know the how, why and when, as well what can go wrong. You are the type of person who gets the new phone then downloads and reads the 500-page manual, so for you I've added THE DETAIL below every exercise. This explains everything you could want to know about the move.

Move: Thrusters

This super high-energy HIIT move works the quads, calves and lower-back muscles, as well the upper back.

The brief

■ Look above you and stretch your arms up high. If the ceiling is less than 1.5m (5ft) above your fingertips, it's too low. Stand with your feet apart and slightly turned outwards and your knees bent, then squat down, placing your hands flat on the floor between your feet.

■ Breathe in and, as you exhale, GO! Drive hard with your legs so you stand up fast while simultaneously throwing your arms above your head and jumping in the air.

■ At the top of the jump, bend your knees to bring your feet in towards your buttocks, if you can.

■ Quickly release and straighten your legs as you come back towards the floor, so that you land with your feet apart and your knees bent to absorb the downward force. Drop quickly back into the squat position and repeat the action for the duration of the interval.

The detail

Everybody looks different when they squat. This is because of variations in the pelvis and femur (the long bone in your thigh), so don't try to copy anybody too closely; simply put your feet at an angle that feels right for you. Going deep into the squat is totally safe when you are working with just bodyweight or relatively light

additional load, such as a medicine ball or dumbbells that weighs less than 20 per cent of you bodyweight.

The most critical stage of this move is the transition when you shift from downward momentum to the upward thrust of the next repetition, when it is very important that you stay in control. If it feels as though gravity is winning then you need to slow down your descent.

Your arms are important: the intensity of this movement is reliant on all the muscles in the upper back and shoulders being involved, so ensure that you get both arms above shoulder height and pointing at the sky rather than in front of you.

The payback
Your heart and lungs will be screaming once you are working at a high intensity. Since this exercise incorporates a deep squat, your gluteus maximus is first going to be dragging you back up to the standing position and then helping to propel you into the air during the jump phase. Your quads, calves and lower back muscles all play a role, but don't forget that your upper back is very important since it assists in the movement of the arms and stabilises your torso.

If this is a new move to you then you will most likely find that you experience some serious DOMS (delayed onset muscle soreness) in your inner thigh muscles. This is mainly due to the work they do helping to decelerate your bodyweight while landing, and while dropping into the squat ready for the next repetition.

Why this made the HIIT list
This exercise is excellent because it works a wide range of muscles. Its impact is further enhanced when the use of weights is incorporated, but it is very important when doing so that you maintain the smooth, integrated pattern of the manoeuvre rather than separating it into different components and end up just doing a squat, jump and shoulder raise.

Move: Gecko

This dynamic HIIT move is all about superb technique at high speed. Get it right and you'll be truly flying.

The brief
■ Get into a press-up position, with your feet wider than hip-width apart, heels raised and your elbows slightly bent.
■ Lift up your right hand above shoulder height and replace it with your right foot. Your chest should be nearly touching your front thigh. This is the 'go' position.
■ Explode upwards into the air by pushing really hard into the front foot and the hand that is on the floor. Switch your hands and feet while you are in the air.
■ You should land with your left foot and right hand on the floor, your right leg back and your left hand in the air.
■ Repeat the move immediately and constantly for the duration of the interval, switching sides each time and ensuring you push up high into the air.

The detail
There are various versions of this move that only require you to lift either your hands or feet – rather than both simultaneously – but just doing these simply won't enable you to reach the required level of intensity. The factor that sometimes holds people back when performing the Gecko isn't strength but more a lack of mobility in the hips and flexibility in the hamstrings (tight shorts don't help either). If you fall within this group then it is worth addressing these issues, and practising the manoeuvre will help.

As fatigue creeps in you'll notice that your chest starts to 'rest' on your front thigh – that's cheating! Keep a space between your chest and leg at all times or you won't be able to use the power in your glutes and lower back to generate enough thrust for the next repetition.

1   **2**

The payback

This move involves load bearing on all four limbs, which brings the added bonus of making the core fire like crazy to stabilise the spine, pelvis and ribcage. The cardio demands are impressive too.

Supporting all your bodyweight throughout the activity offsets the fact that the massive muscles of the quadriceps aren't being asked to propel the mass of your body vertically. Upper-body strength activity is a key feature during the Gecko, with chest, arms and abs all involved in either propulsion or stabilisation (individually or simultaneously).

Why this made the HIIT list

This HIIT move scores a maximum five out of five on account of it involving load bearing on both the upper and lower body. It's fairly intense and definitely hard work, but the benefits are so great it is well worth forcing yourself to do it.

"Upper-body strength activity is a key feature during the Gecko ... "

Move: Run-ups

Every muscle gets worked with this sensational HIIT exercise. Drive forward for big gains.

The brief

■ Begin by running on the spot, starting with low knees and, as your leg-speed increases, bringing up your thigh/knee to just below hip height.

■ Swing your arms so that the hand that is opposite the raised knee is up in the air. Use your arms to generate force and provide a connection between your upper and lower body.

■ In terms of body position, run on the balls of your feet with the heels never touching the floor, and look ahead rather than at the ground.

The detail

Standing with your feet slightly apart means that the action of running on the spot more closely resembles the movement involved with normal running. This is because just lifting your knees up in front – as used in the classic 'get your knees up' method – calls upon the hip flexors rather than requiring the quads, glutes and hamstrings to work. The best way of thinking about this is to picture yourself accelerating forwards rather than simply staying on the spot (if you compare this to a sprint start using starting blocks, you'll know that there is an element of driving from side to side involved with generating forward motion).

It's very important that you stay up on the balls of your feet to generate springiness through the ankle, but also to ensure that the large blood-pumping calf muscle is engaged in the exercise to the maximum.

"It's very important that you stay up on the balls of your feet ... "

1

2

A good sprinting technique is a combination of tension and relaxation, the latter being most evident in the hands, head and face but also, less conspicuously, around your rib cage. Nobody looks good while they are sprinting on the spot, so just get focused on the job in hand and drive your arms hard. Bring your knees up to just above hip height and do everything you can to get as many muscles involved in the action as possible.

Running on the spot and Run-ups can be enhanced by the use of a harness. This goes around your waist, then you 'lean' or 'pull' against it when you run to give the sensation of resistance between you and the ground.

The payback

When you watch footage of a sprinter running in slow motion you'll see that there is a seamless blend of tension and relaxation, as muscles transition between propulsion and the inactive phase between concentric and eccentric contractions. Every muscle in the body is involved in this exercise but, in HIIT terms, the greatest payback will be within the muscles of the lower body.

Why this made the HIIT list

HIIT is defined by the ability to force the body to balance and move between aerobic and anaerobic activity, and sprinting either on the spot or with propulsion is one of best ways of achieving this reaction. Done well, it feels as if your entire body is working harmoniously. Done badly, it looks as if different parts of your body aren't connecting with each other. In summary, this action is the epitome of high-quality human movement.

Move: Plunges

**This HIIT exercise dynamically engages the abductors
for a truly explosive workout. As always, 'learn it and then work it'.**

The brief

- Place one foot quite far out in front of you and bend that knee to almost 90 degrees, making sure it doesn't go forwards of your front foot, so your thigh is parallel to the ground.
- Bend the back leg so there is a 90-degree angle at your knee and lift the back heel off the floor so you are on your toes. You should feel a stretch through the hip on the back-leg side.
- Position your arms, elbows bent, as though you are about to start a sprint, with the one on the back-leg side raised and the other by your side.
- The plunge action requires you to momentarily drop down lower into your lunge, then jump up in the air while simultaneously switching legs, so you land with your back foot forwards and vice versa. Pump your arms, both to help you to balance and to propel you upwards by utilising the contralateral force generated between the upper and lower body.
- Immediately repeat the action for the duration of the interval, so that each repetition blends into the next and there is absolutely no pause between each part of the movement.

The detail

A 'plunge' is a portmanteau word that describes an action that combines a 'lunge' with a 'plyometric' phase during which you swap legs in the air. This is one of those exercises that really becomes clear when watched in slow motion or broken down into stages and practised step by step. It is important that every part is carried out correctly; it's all too easy to cheat by simply skimming the ground rather than propelling yourself up and down. In real time, everything happens so quickly that is not worth trying to get your conscious brain to think about what the left and right sides of your body are doing, which is why you should get a feel for each part slowly beforehand.

The best way of knowing whether or not you are going at a good speed is to count the number of repetitions (leg changes) that you do in a 30-second period. If you doing more than 50 then it is likely that you are not going deep enough into each of the plunges.

Unlike the exercises that mimic running, the movement of the arms during plunges isn't purely backwards and forwards from the shoulders; instead, use the muscle actions from the shoulder, chest, and upper- and lower-back muscle groups to generate power to get your bodyweight off the floor.

Don't hold your breath while performing this move. A small, controlled puff of air out when you start each repetition will give you the mobility around the ribs that is required to enable force to be generated by contralateral movements between the upper and lower body.

1

The payback

You may have noticed a pattern here: propulsion exercises that involve the quads, hamstrings and glutes seem to be great for HIIT. This is because the sheer volume and mass of muscles worked during these moves makes them really impactful. If you embrace the process and throw yourself into the movement then every set should provide enough stimulation to push your body into high-intensity mode. In HIIT terms, this move is virtually perfect: it's contralateral, explosive, fires your core and burns body fat!

Why this made the HIIT list

Many claim that doing squats is the best way to achieve the strongest legs but, while they are good, they rarely involve any explosive action. A plunge, however, works deep in the adductors and dynamically engages the abductors, while also having the desirable and often-overlooked benefit of 'toughening up' the ligaments and tendons of the muscles and joints involved. This occurs as a result of the body's need to manage acceleration and deceleration forces.

Move: Big Jump Jacks

This enhanced version of a HIIT classic gets the blood pumping and gives the body a complete workout.

The brief

■ Start with your feet in a neutral position, knees bent and your hands just touching the top of your thighs. This is the starting position.

■ Jump upwards and forwards. You are aiming to land 1m (1ft) in front of your starting point with your feet wide apart and knees bent.

■ Now reverse the action and jump backwards, throwing your arms upwards and backwards at the same time.

The detail

Could there be a more inclusive, integrated exercise than the Jumping Jack? Loved by aerobic teachers and army PT instructors alike, it really is a fantastically simple yet intense exercise. A regular Jack is when you jump out and in and flap your arms to at least shoulder height. For extra HIIT impact, this move has been designed with an added jumping motion and more dynamic arm movements.

The very best version of this will have you travelling forwards more than approximately 50cm (2ft), as if you are doing an athletics-style long jump from a standing start. The secret of this movement is to think about

your hips. Some of the forward motion is generated from the action of the arms, but it is mostly those great big gluteus maximus muscles that are moving your pelvis from A to B. This is why most people find the 'return' phase of the movement (the jump backwards) to be hardest part, because the combination of the arms swinging backwards and not having those big

glute muscles to initiate the force means that you have a lot less muscle to call upon.

Another common difficulty is that your shoulder joints get 'jammed', which can occur if you don't follow the correct technique: when your arms are down below your shoulder your hands should be pointing flat to the floor, but as your arms rise above shoulder height your arms should rotate. The easiest way to think about doing this is to start with both thumbs pointing towards each other, then as the shoulders rotate, point the thumbs first upwards and then behind you.

The payback

If you observe people doing regular Jumping Jacks you'll notice that it seems possible to settle down into a rhythm, the movement gets smaller and before you know it you are cruising rather than pushing yourself. The big component of this move makes all the difference: regular Jacks don't place much demand upon the lower- or mid-back, spine or shoulder girdle – the big version does. This is because throwing the arms back as you raise them involves scapular retraction, which is fantastic for your posture and conditions the muscles that function between each rib, and you also dynamically extend the spine, which brings the added bonus of working the super portion of your gluteus maximus. Also, in comparison to regular Jacks, this move calls upon and develops the fast-twitch muscle fibres during both the launch and landing phases.

Why this made the HIIT list

This move calls upon the biggest and most resilient muscles in the body that we need to be involved to invoke high intensity, while at the same time integrating all of the invisible deep-torso 'hero' muscles that manage stabilisation of the trunk and the transfer of force throughout the body.

Move: The Rabbit

**The Rabbit works the chest, arms, shoulders, hips and legs ...
Phew, it's exhausting just listing the body areas this move impacts.**

The brief

Warning – if you usually use a mat to protect your hands during floor work, don't use it for this move as you will either catch your feet on it or find that it slides from underneath you, causing you to crash on to your face on the floor.

■ The starting position is best described as being 'strong and ready'. Kneel down on your hands and knees, with your hands under but spaced slightly wider apart than your shoulders, elbows bent and your weight in the heel of your hands rather than on the fingers, which are spread. The feet are spaced wider apart than your regular standing width. Lift your knees off the floor slightly so that your buttocks are level with or just below your shoulders.

■ Jump your feet forwards so that your knees move outside and close to your elbows, putting all your bodyweight on your hands and keeping your feet off the floor.

■ Reverse the action so that your feet return to their starting position on the floor behind you. Experienced athletes will recognise this as being similar to a Squat Thrust. The difference between them is that for a Squat Thrust your feet will touch the floor when you jump both forward and backwards, while in the much harder Rabbit version they only touch the ground when your legs are extended behind you.

1

2

The detail

This move is clever and fun to do. While it looks like it's a predominantly lower-body exercise involving moving the legs forwards, there is some unexpected upper-body action focused on the chest, shoulders, arms and abdominals. As you initiate the jump, you'll probably hold your breath for a split second (that's fine), then your pecs, anterior deltoids and biceps will all contract to make your hands grip and arms brace (tense) your upper body. Your hip flexors are the prime movers of the legs for the forward part of the move, then as your feet hit the floor at the end of the rep, the quads, hamstrings and glutes all kick in.

The biggest challenge posed by this move is achieving sufficient speed – you have to really commit yourself to it or else your feet will just drop to the floor. The first time you try it you'll most likely be able to do multiple repetitions but find that they are punctuated by gaps. If that is the case, you have to speed up because you are doing a good but not high-intensity Squat Thrust rather than the Rabbit.

One of the key benefits of this move is that you are holding your weight in your arms, chest and shoulders, so you have the added bonus of developing strength and endurance while simultaneously placing stress (in a good way) upon your cardio-respiratory system.

The payback

Your chest and arms will be worked really hard throughout this move, which is why it gets a high HIIT score. Another benefit is that you don't experience the unpleasant (and unproductive) impact on the wrists that occurs during some quadruped movements (ones involving four points of contact with the ground). Your core is also switched on and required to adapt to the demands of the move. For me, core is all about the ability to manage and transfer force throughout the body and this move shifts the stability demands to the upper and lower body not once but twice during each repetition.

Why this made the HIIT list

For achieving an impressive high-intensity response, this really is one of the super-hero HIIT moves since it stimulates the body in an effective yet not-unpleasant way, unlike some moves. The first time you try the Rabbit, you may get to 5 seconds and think everything is fine, but by 15 seconds you'll be begging for it to end. Push through that feeling and continue to the end of the set; it's well worth it since this is one of the most challenging but productive moves around.

3

4

Move: Bangbang Firing

This truly explosive move massively enhances agilty, balance and body core – and works every part of the body for impressive results.

The brief

■ Stand in a 'braced' position, which means that your stance is so strong and grounded that if someone were to run into you they would canon off you and you'd remain upright. Stand with your feet wide apart, knees slightly bent and leaning slightly forwards. Bend your elbows and raise your hands, shoulder-width apart, to chest height with your palms facing downwards.

■ Jump up in the air and propel yourself so that you rotate 180 degrees, resisting the urge to use your hand and arms to help you twist.

■ Instantly jump back the other way to your starting position, thus avoiding doing a full 360 degrees and making yourself dizzy. Repeat for 30 seconds.

The detail

This move was taught to me by a physiotherapist in the USA who was working with a 2.21m- (7ft 3in-) tall basketball player with a weak core. The player's job was to be fast and agile for just a few seconds at a time – no mean feat when you weigh more than 136kg (300lb) since quickly pivoting that amount of weight isn't an option because he just stuck to the ground. In response to this problem, the jump-turn 'firing' move was born.

It's extremely hard not to use your arms when performing this move, but you must resist the temptation since doing so helps to build up momentum and thus reduces the workload for the core. The 'secret' is to momentarily turn in the 'wrong' direction, as if to wind up your torso ready for action (it's debatable whether this is the torsion or recoil phase of the move once you get past the first repetition, since all the phases will blend into each other). It is also important to exhale when you initiate the rotation by turning the 'wrong' way; if you don't, you'll have so much intra-abdominal pressure that your torso won't be 'fluid' enough to let the upper and lower body disassociate from each other.

What goes on with your hips, knees and feet is also critical for this move. Imagine that you have an elastic band wrapped around your ankles and at knee height, both of which have to be kept stretched tight. If either band becomes loose then you haven't kept your knees and feet wide enough apart during the jump or landed phase of this move. This isn't just a hypothetical suggestion, either: if you have a set of thera-bands or resistance tubes then you can wrap them around your

1

ankles and knees and instantly intensify the manoeuvre by ensuring that they stay taunt throughout the jump, rotation and landing stages.

The payback
Your heart rate will go sky high during this exercise, making it one of the kings of all HIIT moves, although the balance, coordination and muscle-recruitment gains are equally impressive. This is one of those exercises that you have to do for yourself to be able to understand what all the fuss is about, although it is worth noting that the real payback probably won't be instantly recognisable, and you won't notice the benefits until you are required to burst into action in real life rather than during a training session. When this does occur, you'll hopefully feel more agile and capable of controlling your body, and be especially awesome at fast, dramatic changes of direction.

Why this made the HIIT list
Simply because it is a fantastic move that has real-life benefits for agility and speed.

Move: BurpEE (a burpee but bigger)

Power through these super-charged butt-busting BurpEEs for a phenomenal HIIT move that works a mass of muscles.

The brief

■ Look around and above you to ensure you have sufficient space in which to move around without hitting anything. Go down into a squat position, with your feet flat on the floor and hip-width apart, knees bent at 90 degrees, back flat, and your hands on the floor shoulder-width apart. This is the starting position.

■ As soon as they touch the floor, transfer your weight on to your hands and jump your feet backwards. The goal is to fully extend your body into what looks like a press-up position, with your feet shoulder-width apart and your back and head in a straight line.

■ Immediately jump your feet forwards to the starting position, then rapidly stand up and jump in the air.

■ Throw your arms above and behind you while simultaneously doing a hamstring curl so your feet hit your buttocks.

■ As soon as you land, go straight into the next repetition.

The detail

Hardly anybody actually enjoys doing Burpees as they are such hard work, but for HIIT they deliver fantastic results.

There is a fine and subtle line between doing this exercise in the correct or wrong way, so it is worth breaking down the movement into phases. First, the squat from standing: this has to be fast and your heels will lift off the floor as a result of the speed. Next, the transition back to the press-up position: your hands should not be smashed into the floor. If they are, this is because you didn't squat low enough before moving your feet. When you drive your legs backwards, the goal is to land your feet as far away from the hands as possible; you'll know if you have achieved this by assessing how high your hips are – if they are above your shoulders when you are fully extended then there's no way you have positioned the feet correctly.

NB Although you adopt the press-up position, you shouldn't actually incorporate a press-up into the move

since this would kill the intensity (albeit providing a strength challenge to the chest, arms and shoulders).

As soon as you have jumped your feet forwards into the starting position, the vertical jump section begins. This involves driving and generating upward force from both your upper and lower body. The goal of this phase isn't to jump as high as you can – focus instead on making it feel strong.

There's very little time to think about it, but almost as soon as you are in the air you need to start the extension component of the move – literally throwing your arms up and back and quickly flexing your knees as you try to kick your buttocks with your heels. Doing both of these means it is almost guaranteed that you'll arch/extend your spine: this is the goal. Instinct will kick in as you start to drop back down towards earth, so just land naturally with your knees slightly bent, then go straight into the next rep.

The payback

If a good HIIT move is scored on the basis of how many muscles it integrates, the amount of power you need to generate it and the speed at which you perform it, then this super-charged BurpEE ticks all the boxes. It's also a winner because it's not overly complicated and doesn't require too much thinking. This is ideal, since as soon as an exercise demands precision, it loses some of its ability to create intensity; these moves are designed to be an intense workout rather than a test of skill.

Why this made the HIIT list

What 'super-charges' this move is the extension of the spine, shoulders and hips when you are in the air. The inclusion of this aspect does two things: it increases the intensity/challenge of the move, and it stops you from doing the most common Burpee cheat, which is jumping in the air but simultaneously looking at the floor and barely leaving the floor. Try doing this for yourself and you'll see what I mean.

Move: Arrows

Awesome for developing core strength while working the abs, arms and butt ... This move is what HIIT is all about.

The brief

■ Lie face down, move your hands in front of you and place your palms on the floor with your arms bent at right angles at the elbows. Tuck your toes under your feet so that your shins and knees are elevated. This is the starting position.

■ Push your hands and toes against the floor really hard so that your whole body is propelled into the air, keeping your legs, back and head in line so that you retain the push-up position while in the air.

■ Come back to land on your hands and feet, maintaining the push-up position.

■ Lower yourself to the ground and immediately lift your arms, hands, legs, feet and chest off the floor. The only body part on the ground will be your hips.

■ Return to the starting position and immediately go into the next rep.

The detail

The move looks self-explanatory, but to promote it from humble Back Raise to ab-tastic Arrow you have to totally commit to going fast. Going slowly will deliver some really good strength gains but it's the speed that produces the responses we want in order to be able to call it HIITastic.

The key difference between a Back Raise and an Arrow is that the push phase of the latter is just that, 'a push', whereas a basic Back Raise is a pull. When you start the push phase you need to think about bracing your torso so that you connect the upper and lower body together. This move won't feel like any other press-up you have ever done; it works the chest, arms and shoulder, just like regular push-ups, but the demands placed on your abs are magnified enormously and that hits your chest in the same way

that performing a heavy pull-over with a barbell would, but with the added bonus of your legs being part of the action.

On the subject of legs, their role in this exercise is to get the front of your body off the ground. If you overwork the leg component you'll find that your hips shoot up in the air and that you end up doing a pike rather than an Arrow. As you drive upwards, the highest you'll get off the floor is about the same depth as your body – 30–40cm (12–15in) – so if you find yourself in a full press-up position and don't gain any elevation then something went wrong and you probably started with your hands too close to your head.

As you hit the ground (competing against gravity makes this phase of the exercise too slow to be HIIT) you want to go instantly into reverse by thrusting your arms up and back and lifting your feet, knees and thighs off the ground (the higher the better). Don't 'hold' this extended position, go straight into the next rep.

The payback
Abs, arms, buttocks, heart, lungs and core will all be on fire after a set of these Arrows. Moreover, the move really stimulates the body because the moment you push your hands into the floor you'll be creating intra-abdominal pressure (which is awesome for developing strength in the core), and because the arms are in effect 'above' your head when you push against the floor you'll be hitting your chest muscles while maintaining scapular retraction. Moves rarely achieve this, and the conditioning is great for posture.

Why this made the HIIT list
This move is borderline nasty so it almost didn't make the cut, but, despite the fact that when you become fatigued you're likely to hit the ground with a thud, I think the pros offset the cons.

Move: Skater

Glide your way to a toned, lean body thanks to this highly demanding HIIT exercise's superb plyometric action.

The brief

■ Look around you and ensure you have plenty of space in which to glide sideways. The goal is to go wide without compromising speed and depth.

■ Start in a half-squat position with your feet shoulder-width apart, knees partially bent and your arms held slightly away from your sides.

■ Jump sideways, landing on your outside foot, with both knees bent and your inside knee off the floor and the inside foot positioned near the bent knee of the standing leg. The goal is to 'catch' your bodyweight on the bent outside leg and instantly drop down to a half-squat position.

■ Swing your arms in the direction of the jump, bringing the inside arm across the mid-line of your body and swinging your outside arm backwards. The motion helps to create some of the force that propels you sideways, and the shape you make upon landing should resemble that made by an ice-skater in action.

■ When you hit the deepest point of the single-legged half squat, jump or drive sideways in the opposite direction, using the arms to generate lateral force.

■ As soon as you have landed and reached the deepest part of the half squat, drive again. Repeat the action for the duration of the interval.

The detail

If your coordination is poor or average then you need to start with small movements or you'll be in trouble and lose technique as soon as fatigue sets in. The first jump (glide phase) that you do is usually bigger than the ones that follow due to the fact that you are going from a standing start rather than trying to go against the momentum of the previous rep.

When you jump, there is a temptation to turn your head to look at your foot. Try to avoid this since it will make your torso rotate; the goal is to keep your shoulders pointing straight ahead. This has the

beneficial knock-on effect of keeping your hips facing forwards and thus minimising the amount of adverse rotational forces being inflicted on your knees.

This move should be done fast; if it looks as if you are pausing during each rep then you are probably going too wide, too low or too slow. You can fix this by limiting the size of your glide/jump. A good guide is to limit the total width of your jump/glide to the length of your body, and aim for a rhythm of 60-plus reps per minute.

1

The reason the jump phase is referred to as a 'glide' is because the move is inspired by speed skating (in particular, the sprint-start part of the action), although when doing this as an exercise you don't lean forwards as much as a real skater, who is simultaneously pushing backwards as well as sideways to generate forward motion. That's great for intensity but only doable if you have a huge amount of space.

Your arms, chest and shoulder movements are significant as they produce about 30 per cent of the lateral force. It is therefore important to ensure the arm action isn't just a side-to-side swing, which helps with your balance, but is actually focused and aggressive. This motion contributes to the HIIT effect thanks to something called PHA (peripheral heart action), which means that your body has to work really hard to spread the available blood around your system.

The payback

This exercise is a fantastic full-body workout thanks to its three-way power/plyometric action, so is well worth the funny looks you may get if you do it in a gym where there is a focus on pumping iron and looking hench. Ignore the posers: they don't know what wonderful things the move is achieving.

Why this made the HIIT list

The Skater is one of the few moves that incorporate powerful lateral forces (in the jump/glide phase), so even if you were to ignore the substantial cardiovascular and energy-burning benefits, it's recruiting more muscles than almost any other HIIT around.

BODYWEIGHT HIIT EXERCISES INTENSIFIED WITH DUMBBELLS

Move: Bottom Up

Grab a pair of dumbells and tackle this highly rewarding HIIT move that works the arms, upper back and legs.

The brief

■ Holding the dumbbells, start in a press-up position but with your bottom up higher than usual, in what is called a pike position, on your toes, with your back flat. This is the starting position.

■ Using a jumping motion, move your feet towards your hands. As soon as they touch the floor, begin to stand up, simultaneously picking up the dumbbells.

■ Keep the dumbbells close to your torso until they reach shoulder height, then push them above your head.

■ As soon as you reach maximum height, reverse the move until the dumbbells touch the floor again. Jump your feet back to the starting position and do it again. It's worth mentioning that you should think about the floor surface on which you are doing this exercise; if it is wooden or made from any other material that could mark or be easily dented then you need to protect it with a mat before you start.

The detail

The reason the buttocks are up in the air is because the lift and drop stimulates more muscle activity and intensity than keeping them low would (in a regular Plank or Press-up the hips usually stay below shoulder height). When you jump your feet forwards you must 'commit' to making the jump big – your toes should land in line with your hands/dumbbells. That said, since the move is very intense even when your feet stay on the floor, don't feel obliged to leave the ground.

The lifting of the dumbbells is achieved as much by muscles in your upper back as it is by your arms. This means that the dumbbells go through an arc, rather than being lifted in a straight line from the floor to the ceiling. The down phase of the move is of course somewhat assisted by gravity, but if you are doing it properly you should be forcing the dumbbells and your body to get back to the ground as quickly as possible, rather than letting gravity do all the work.

The payback

In the simplest interpretation of HIIT, the more muscles, the bigger the range of motion and the faster you move your body, the greater the benefits/challenge will be. Based on these parameters, this move is one of the most rewarding uses of your time.

Why it made the HIIT list

You can do HIIT to burn calories, strengthen your cardiovascular system, and become stronger, faster and more agile. However, it is rare for one exercise to achieve all of these outcomes, but that's exactly what Bottom Up does brilliantly.

"The dumbbells go through an arc, rather than being lifted in a straight line ... "

4

3

Move: Split and Rip

You really have to push yourself for this demanding HIIT exercise. You'll be rewarded with a move that brilliantly majors on working the lower body.

The brief

■ Hold both dumbbells and get into a lunge position with your left foot in front and the right dumbbell slightly lower than the left. Bend both legs so that the right dumbbell goes close to the floor, but doesn't touch it. This is your starting position.

■ Jump in the air and lift the right dumbbell straight up and above your head while simultaneously switching legs.

■ Once you have landed, lower the dumbbell so you are in the starting position and repeat the process with the left hand taking the lead.

The detail

Holding a weight in your hand stimulates your core to engage. This should help you to balance and stay upright, but since this is a challenging exercise you should try to look ahead, not at the floor, to further help you keep your balance.

The lunge itself is a plyometric move, which means that the landing phase is as important as the jump phase. As soon as you feel your feet touch the floor you should consciously sink lower into the lunge, as if you are absorbing the impact, rather than stopping dead. This will make the inner muscles in your thighs, glutes and hamstrings scream. If you can work through this then you will come out the other end stronger and more agile.

The dumbbell part of this move should be fast and rhythmic, so you only need to use a medium weight. You will know if you have got this right if your jumps, lunges and lifts look and feel smooth and sequential. If the move is punctuated (stop–start) then it is likely that the weight is too heavy.

The payback

This isn't the kind of move that you do for fun. In addition to the spectacular cardiovascular effort and levels of fuel consumption involved, it also brilliantly trains your body through all three planes of motion, engaging your core and getting your lower body to cope with plyometric forces.

Why this made the HIIT list

While this exercise does require a medium to good level of coordination, the rewards are worth the challenge. It will take you a few tries to perfect the movement, but it isn't complicated for the sake of it; every element adds up so the entire move is worth more than the sum of its parts.

Move: Double Thruster

**This move is all about strength and conditioning.
Work on your technique for big HIIT gains.**

The brief
■ Hold your dumbbells at chest height and at right angles to your body with your elbows bent, standing in a sumo-squat position with your feet wide apart and knees slightly bent.

■ Drop down until your thighs are parallel to the floor, then lean back so that the dumbbells feel as if they are above your hips rather than in front of you. This is your starting position.

■ Push your feet hard into the floor and simultaneously drive the dumbbells straight upwards. The force from your legs may make you take off but it's not compulsory.

■ Once in the air, try to straighten your legs and point your toes, but keep your legs and feet wide so that you land in the same position as when you took off. As gravity takes over, pull the dumbbells back to chest height so you are ready for the next repetition as soon as you land.

The detail
The most common mistake with this move is that after the first repetition the range in movement/depth of the squat diminishes significantly; it requires a lot of discipline to maintain the technique. Doing the exercise in front of a mirror makes all the difference as it allows you to check and monitor your technique throughout the set.

The most subtle element of this move is that you are leaning backwards in the starting position. This means that your abdominals have to work very hard to stabilise your torso, when all you really want to do is lean forwards from the hips and let your glutes take over. The easiest way to maintain good technique is to lift your chin and look upwards rather than forwards when you are in the up phase.

1

"The easiest way to maintain good technique is to lift your chin and look upwards ... "

The payback

This is a super-high-intensity exercise thanks to the huge demands that are placed on your pelvic floor and all the other muscles in the pelvis area throughout the move. These muscles are not cosmetic, in fact you can't even see them, but strengthening and conditioning them produces countless knock-on effects that benefit you in everyday life and during athletic activities.

Why this made the HIIT list

For an exercise that only requires a little bit of space and a pair of dumbbells, this move delivers very impressive cosmetic, physiological and athletic gains.

Move: Bell Ringer

The Bell Ringer is a fantastic HIIT move that gives the lower back and abdominals a highly impressive workout.

The brief

■ Use your lightest dumbbells for this move. Stand with your feet hip-width apart, knees soft, then bend forwards from your hips (the starting position in the photograph is a classic hip hinge) and let your arms hang straight down.

■ Swing both dumbbells sideways and above your head, turning your head in the direction of the swing. If you can, simultaneously jump in the air (the lift-off is really a result of you creating a decent level of momentum with the dumbbells as much as you powering upwards with your legs).

■ As soon as you start to fall back to the ground, force the dumbbells back through the arc and repeat on the other side.

The detail

Without supervision, many people get this exercise wrong even before they start to move. This is because there is a tendency to bow forwards through the spine rather than bending forwards from the hips. Only the very lowest joints in your spine should be flexing and, if you do it correctly, you should be able to maintain the natural concave curve in your lower back.

1

2

The second-biggest mistake is that the dumbbells are propelled using arm and shoulder muscles rather than as a result of force generated from the torso. Your goal is to keep both arms straight throughout the whole movement and make the arc through which the dumbbells travel as big as possible.

Breathing is a bit tricky during this exercise because in a hip-hinge position your instinct is to create intra-abdominal pressure by holding your breath. However, breath-holding and HIIT activity don't go together, so exhaling on the lift and sucking air in on the down phase is the best method.

The payback

The Bell Ringer generates less of a cardiac response than some of the other exercises, but the sheer volume of activity that occurs in your abdominals and lower back offsets this and makes it well worth doing.

Why this made the HIIT list

It might make you look a bit silly, it doesn't include plyometrics, and no you won't enjoy doing it, but this movement has a fantastic effect upon your physical strength and dexterity.

Move: Swing and Step

**This superb body-sculpting and heart-pumping move
is popular with professional athletes and is great for HIIT.**

The brief

- Start in a drop-lunge position with your right foot forwards and your knee bent, then position both dumbbells in front of your left thigh, ready to swing them over to the left.
- Jump up high and simultaneously switch legs, while also swinging the dumbbells in an arc to the left.
- Land back in a drop-lunge position with your left foot forwards and the weights over your right thigh, ready to repeat the action. As you feel your feet touch the floor, bodyweight momentarily keeps dropping towards the floor so you have a moment of plyometric activity.

The detail

Before you begin, you need to commit to making the leg action a dynamic jump rather than letting your feet simply slide on the floor. The dumbbells you are using should be light to medium in weight since the arm action needs to be as fast as possible. In fact, the movement of the arms dictates the speed of the entire movement.

1

2

3

Since this exercise starts and finishes in a split stance, unlike a squat, the front leg works harder than the back one. The biggest mistake you can make is to try to do a lunge that is too long; keep the distance between your feet to approximately half of your overall height.

The payback
If you get this move right, your heart will be pounding on the inside of your ribs like a bass drum. Sprinters and other high-performance athletes use similar moves (usually without the dumbbells) to encourage leg and arm speed. As our goal is to make everything high intensity, this move is ideal since the dumbbells magnify the energy demands and cardiac response without requiring any additional coordination or skill.

Why this made the HIIT list
Sprinters display both highly desirable levels of performance and enviable physiques – enough said.

4

Move: Stones Lift

Subtle yet highly productive, this move combines an impressive cardio workout with a great lifting action.

The brief

■ This move goes from left to right as well as up and down. Kneel on your right knee with your left leg bent to 90 degrees. The dumbbells should be positioned on the floor just inside your left foot.

■ Stand up fast using the strength in both legs equally and try to 'leave the weights behind' so that they don't start their journey until the legs are almost straight.

■ At this point, bring up the weights in an arc above your head. You aren't trying to jump in the air, however, you need to at least lighten the pressure on your feet or else they will stay stuck in the start position.

■ Switch your feet while in the air and come down to land in a squat with your right knee on the floor and your left leg bent at 90 degrees, driving your arms in a downward arc.

The detail

When I see people doing this move with their back rounded at the outset I am reminded of the marble sculpture the Farnese Atlas, which depicts the Greek god holding the celestial spheres on his shoulders – it looks awkward. You should instead keep your back straight.

If you have various dumbbells, this move should be performed with your heavier option. This is because although the weight makes Stones Lift slow compared to most of the other HIIT moves, it is very intense thanks to the fact that moving the weights in an arc magnifies how heavy they feel (by as much as two or three times the actual weight of the dumbbells).

Every muscle in your body is activated to help perform this move. When the weights reach the highest point of the arc you may feel as if your feet are 'unloaded' and that you could jump in the air. Don't: the benefits of rushing your hands back down towards the start of the next repetition outweigh those of adding a jump.

The payback

During the first few reps you may not notice too much fatigue, but 5–10 seconds into the set you'll realise that the stress being placed on your cardiovascular system is significant. This, coupled with the huge demands placed on the leg muscles to get your entire mass both lifting and changing direction, makes this move subtle but highly productive.

Why this made the HIIT list

Although Stones Lift has a slower tempo than most HIIT exercises, it's more than capable of elevating the heart rate to the desired levels to make it an important HIIT move.

Move: V-jump

The V-Jump activates the posterior chain of muscles for an unbelievably intense workout.

The brief

■ This is a double dumbbell move and is best performed using two lighter weights; if you only have a medium/heavy set you can, however, adapt it and hold just one dumbbell with both hands. Stand your feet hip-width apart, knees soft and your back straight. Hold the weights level with or just below your crotch with your arms straight.

■ Jump forwards while simultaneously swinging the weights above your head, you should land with your feet apart. When you do land the weights are at their highest point, immediately reverse the movement so that you jump your feet back together and pull the weights back down to their start position.

The best way to determine how wide your feet should be is to stand with your feet together and your arms outstretched at shoulder height, 'Vitruvian Man'-style. The subsequent width between your elbows represents the perfect distance that should be between your feet as you land.

The detail

Positioning the weight just below your crotch will probably cause you to lean forwards slightly. Try to ensure that you return to this starting position at the end of every repetition or else the arm action will diminish in power.

This move requires a lot of control in your shoulder area. When you swing a weight in an arc it feels three to four times heavier than it actually is. At the highest point, the dumbbell just wants to keep going and complete its circle, so you are going to have to activate muscles in your chest and shoulders to decelerate the dumbbell and reverse its direction of travel in a split second.

The jump forwards is relatively easy because the weights should be dragging you in that direction. The skill and challenge predominantly occurs when you

land with your feet apart and, as when performing many of these hip moves, you need your movements to be fluid and to recoil during this eccentric phase. To achieve this, as soon as your feet touch the floor sink low into a sumo squat then spring back to the starting position.

1

The payback

Many HIIT exercises prioritise the chest and abdominal areas so it's fantastic to have a move in your exercise portfolio that activates your posterior chain of muscles so intensely. One of the most neglected areas of training is improving your body's ability to decelerate and change direction. This is because so many activities rely upon momentum. The V-jump is different because although it requires huge amounts of thrust, there isn't enough time for momentum to build up before you go into reverse.

Why this made the HIIT list

Beyond the fact that this move is most definitely an intense HIIT exercise, it has the added bonus of being an antidote to the rounded shoulders that are so prevalent among a population that spends too much time sitting and looking at screens.

Move: Shot-put Driver

Take your time and perfect the technique for this legendary HIIT exercise. When executed correctly it works the upper and lower body to brilliant effect.

The brief

■ Hold a medium/heavy dumbbell in your right hand at chest height, then place your right foot in front of you and drop down until your front knee is almost at 90 degrees. Lift your left heel from the floor. Bend your left arm at the elbow and position that hand under the elbow of your right arm. This is the starting position.

■ Quickly step back with your left foot and immediately turn your body through 180 degrees while driving the dumbbell to arm's length to a position high above where your left shoulder was in the starting position.

■ Reverse the action and return to the starting position, ready for the next rep.

The detail

Of all the moves in this book, this one requires the most balance and coordination. Timing the leg and arm action so that they complement each other can be practised at slower speeds and without the weight (learn it, then work it). Once you are happy with your technique, introduce the dumbbell and gradually build up your speed.

Remember, when you push the dumbbell above you it is being driven up by your leg muscles as well as those of your arm and chest so it's going to be moving with considerably more force than it does during a regular shoulder press. It is important, therefore, to be prepared to reverse the momentum you have created when you recoil back to the starting position.

The payback

The Shot-put Driver has everything that makes an exercise great: it demands power, stability, coordination, a connection between the upper and lower body and, my favourite, a tri-planar movement.

Why this made the HIIT list

The shot put is field event performed by Olympic athletes. However, they only have to do one repetition at a time before having a rest, so it made sense to adapt the technique slightly to create an impactful HIIT that can be repeated numerous times without pause.

"The Shot-put Driver has everything that makes an exercise great ... "

Move: Turk Thruster

This explosive move will leave you and your muscles feeling pumped. Combining balance, coordination and strength, it's the perfect HIIT move.

The brief

■ Sit down on your butt with your knees bent and your feet shoulder-width apart and flat on the floor. Bend your right knee slightly more than your left one. Lean backwards slightly, holding the dumbbell in your right hand, supporting yourself on your left hand with your left arm straight and your fingers pointing away from your body. This is the starting position.

■ Push up through your arms and legs, lifting up everything: the right arm and the dumbbell, your hips and your torso. Come up on to your toes if you can.

■ When you can't get any higher, reverse the movement until you momentarily touch the floor with your buttocks, then start the next repetition.

The detail

This move is very loosely based on the Turkish Get-up (TGU) that is performed with a kettle bell, though I would argue that this version is better because the TGU is usually punctuated into various sections. It is a very primal exercise, a phrase that is used in the fitness industry to describe a movement that's a very natural caveman movement, but which is exaggerated to increase the challenge.

If you are tight in your chest muscles then the hand position on the floor might be challenging. If this is the case then you need to improve your mobility since this is a position that everyone should be able to achieve.

People of different heights and shapes look very different when they do Turk Thrusters, so instead of comparing yourself to others, focus on finishing with your hips, hand, chest and head as high in the air as you can, then recoil quickly but without smashing your buttocks into the floor. In order for this fantastic exercise to achieve high-intensity status you must go fast, so make sure you don't start showboating – get up, get down, get it done.

The payback

Your arms are going to be 'pumped' after a set of these – not so much the one that's pushing the dumbbell but more the one that's left behind on the floor. The more bend and flex you make it do, the better.

Why this made the HIIT list

Awesome for the core, fantastic for usable functional strength and with the added bonus of really working your arms as part of the process – what's not to love!

3

"Focus on finishing with your hips, hand, chest and head as high in the air as you can … "

Move: Halo

Make big circles for big results. This move works the arms, legs and butt, not forgetting the heart and lungs.

The brief

■ Stand with a wide stance and knees slightly bent. Hold a dumbbell with both hands at hip level and positioned midway between your feet.

■ Turn your body slightly to the left, bending deeply into your left knee and less so into your right one, flexing your hips. Grip the dumbbell with both hands and bring it down above your right thigh.

■ Swing the dumbbell in a circle over your left shoulder while shifting most of your weight on to your left foot and twisting your body further to the left.

■ As the weight travels around the back of your head, shift your bodyweight on to the right leg.

■ As the dumbbell comes over and below your right shoulder, bend your legs so that the weight drops down to the starting position (but don't rest it on the floor).

■ Loop together four or five revolutions and then decelerate and repeat in the opposite direction.

The detail

It's an over-used phrase, but this move really does 'work every muscle in the body'. While the weight of the dumbbell is obviously an important factor, its real purpose is to make you focus on revolving through the biggest feasible circular movement while maintaining

speed (you can make bigger circles by holding the dumbbell further from your torso, though the lack of speed will diminish the effectiveness of the move). That said, despite the fact that the aim is to go fast, it's best to rehearse the move at a modest speed before going flat out.

The mistake that most people make is incorrect leg action, whereby they perform more of a squat movement rather than the shifting their weight from the left to the right side. Done correctly, not only does this shift add significant intensity to the exercise but it also multiplies the number of muscles involved.

The payback

Tri-planar movements are always worth more than the sum of their parts, and this one is even better than those that 'simply' incorporate all three planes – sagittal, frontal and transverse – because it makes additional demands, such as speed, balance and coordination. Moreover, your heart and lungs will be constantly adapting and getting stronger during every rep.

Why this made the HIIT list

It's tough, it's very effective and it's my own creation, meaning it stands apart from the generic and sometimes outdated basic movements that can be found on the internet or are taught in some classes or training sessions.

"Your heart and lungs will be constantly adapting and getting stronger ... "

HIIT EXERCISES FOR THE GYM, GARAGE OR CROSSFIT™ BOX

Move: Squat Rack Drop and Row Combo

Attack this move at speed for an astonishingly intense upper-body

The brief

■ Stand with your feet shoulder-width apart and your toes under the bar. Bend your knees and flatten your back. Grip the bar with the fronts of your hands facing forwards.

■ In one single flowing moment, 'clean' or lift the bar to chest height, pulling it up and into your chest with your elbows bent so your forearms are parallel to the floor and the fronts of your hands are still facing forwards. Straighten your legs.

■ Pivot your forearms and wrists so they are perpendicular to the floor and the palms of your hands and your curled fingers are facing upwards.

■ Keeping your hands and arms in the same position, perform a front squat.

■ Reverse the row/deadlift phase while in the squat, squeezing your elbows back from in front of the bar, then release the weight of the bar so it drops towards the floor.

■ Let the weight bump on the floor, then immediately repeat the move, building up speed but trying not to sacrifice the quality of movement.

The detail

The lift phase must be performed smoothly, but once you get the bar past your knees you can be more aggressive. The goal is to lift the bar to chest height as fast as possible by taking it in a straight line from the floor to your chest, pushing your elbows slightly forwards of the bar.

The descent into the squat starts as soon as the bar has been 'racked' (when your forearms are perpendicular to the floor). Because the squat is fast, you'll be dropping your hips below knee height so the weight you select needs to be modest or else you'll find yourself sitting on the floor with the bar on your lap.

The payback

The Deadlift is often cited as a 'super move' that everyone should do, but it can feel rather 'staged'. Squat Rack Drop and Row Combo, however, frees you up to attack the exercise at speed and to develop functional strength rather than the futile version thereof that is the result of non-compound movements. Layer that functional-strength development with the heightened intensity and upper-body pull phase that this move involves, and you can see why it delivers such useful challenges for the posterior chain of muscles.

Why this made the HIIT list

This isn't Olympic lifting, which is a sport; this move is all about intensity, overload and encouraging the body to adapt. Nor is it Crossfit™, despite the fact that it involves a collection of moves that you'll see laced together in that environment. It's here because the balance between the skill required and the gains acquired are closely matched (modest skill/ impressive gains).

Move: TRX® Pike Pull and Y-fly

This HIITastic offering guarantees some awesome core strength gains and delivers a truly chest-bursting workout.

The brief

■ Face the anchor, standing far enough away for the TRX® to be tight when you are suspended with straight arms. Stand with your feet hip-width apart and your legs straight. This is the starting position.

■ Maintaining strict control and keeping your legs straight, drop your hips/buttocks towards the floor until you are in the pike position and your buttocks nearly touch the ground.

■ Keeping your arms straight and using your shoulders and upper back to initiate the momentum, thrust your hips upwards and forwards. As your arms reach chest height, lift them above shoulder height to finish the move on your toes and with your body and arms making a 'Y' shape.

■ Reverse the movement, dropping quickly into the pike position before pulling yourself up back to the starting position and repeating the movement.

The detail

As you lower yourself into the starting position there should no slack in the straps; think of the straight straps as an extension of your straight arms. When you move into the pike position, your head will fall between your arms as if you are diving into a swimming pool (your hands are in an over-grip position). When in the pike, don't attempt to keep your feet flat – let the fronts lift and grip with your heels.

The up phase (pull) is initiated by the thrust of your hips forwards and upwards while you simultaneously perform a move that's similar to a machine-based Lateral Pull-down, except that only your arms are straight. Once the feet, hips and shoulders are all forming a straight line it's time to raise your arms above shoulder height into the 'Y' fly position.

Ordinarily, I would say that letting the straps lose tension is a bad thing, but for this move you are pulling so hard and moving so quickly that there is a good

1

chance there'll be a moment at the very top of the action when everything transitions between concentric and eccentric muscle actions and the straps go slightly slack. This is fine, just don't pause at the top waiting for something to happen – shift your weight and throw yourself into the next rep.

2

3

The payback

Core, core and more core, plus awesome strength gains for the posterior chain and, thanks to the massive demands made by shifting between the upper and lower body, the heart rate and respiratory cycles are rapid.

Why this made the HIIT list

When people work alone to figure out what to do with a TRX® rather than with some prior knowledge or the help of a PT, they generally end up doing moves that mimic exercises they already know (chest presses, squats, etc.). This killer compound move has been included to show you how to ignite the HIIT process and provide some variety from the familiar squat-inspired movements.

Move: TRX® Skydive
(Sumo Squat – Shoulder Press Combo)

**The potential gains delivered by this move are HUGE. Prepare
for an instense and highly productive workout that truly delivers.**

The brief

■ Stand with your back to the anchor with your feet
wide apart. Hold the handles at shoulder height with
your elbows bent and palms facing down, and lean
forwards until the straps are tight and you feel tension
in your torso. Quickly squat down, going on to your
toes and keeping your elbows bent. This is the starting
position.

■ On the bounce when you begin to stand up, press
your hands upwards and forwards to perform the
shoulder-press section of the move.

■ When you reach the point at which your legs, torso
and arms are straight, reverse the movement to come
back down into the squat, then bounce upwards again
and repeat.

■ As you become confident doing this move, speed
up everything to the extent that you find that you are
performing the up phase of the squat so powerfully that
you leave the ground (it is a jump, but I'd rather the
elevation were a result of your speed rather than being
an intentional leap).

1

The detail

It's easy to overestimate your ability and strength when
getting into the starting position for this move. You'll
probably find that the first time you try it you'll be able
to hold your body at an angle similar to the one shown
in the photographs, but when you are pushing your
hands up at the bottom of the squat position you'll be
shocked by the load on your upper body – so practise
first and 'learn it then work it'.

The sumo position has been chosen because it
helps the exercise to work the whole body rather than
prioritising the upper-body section. In addition, the
amount of glute activation and inner-thigh action
is significant.

Your heels should be lifted throughout the entire move.
As you can see in the photograph, at the highest point
of the move there is an imaginary line running from the
ankle to knee, hip, shoulder, elbow and hand. If you
can't maintain this line, move your feet away from the
anchor point and you should find things become a little
more achievable.

The shoulder-press phase simply involves driving
your arms upwards and forwards so the hands take the
shortest route up until the arms are fully extended (this
sounds easier than it actually is). The 'jump' isn't a
jump, it's a consequence of you driving up hard and
fast, which necessitates flowing movement. When I

2

3

coach this move the most common correction I need to make is to stop people from breaking it down into sections. Instead, smooth out all the pauses and make it one long sequential action.

The payback

A huge amount of muscle activity occurs during this exercise; muscles are either under tension in concentric/eccentric phases or firing isometrically to maintain stability of the skeleton. All of this makes it a win–win exercise for a HIIT session, when optimising productivity during a set time frame is the goal.

Why this made the HIIT list

This is a brutal move, but it's not nasty or complex just for the sake of it; the benefits to core strength are immeasurable. Moreover, when compared to a similar combination of squat and press performed vertically and un-suspended with a bar bell, it engages many more muscles.

Move:
Hang Tough Combo

This muscle-pumping mash-up works every part of the body, making this a sensational HIIT move that truly delivers.

The brief
■ Stand underneath a bar, then look up and move until you are slightly behind it so that when you jump you lead with your hands and chest forwards rather than headfirst.
■ Quickly squat down on the floor and hold your arms out straight and pointing downwards. This is a 'messy' squat rather than the clean version you'd do with a bar bell, so if your hands touch the floor you are doing well.
■ Drive upwards to reach above you and catch the bar, using the momentum from the jump to pull your head above your hands and bring your chest to hand height.
■ Let go and drop straight into the next rep/squat.

The detail
This move isn't a classic exercise – it's a mash-up of several, and therefore it is likely that everyone is going to have their own interpretation of what doing it well looks like. That said, it generally works best when there's no separation between the squat, jump and pull phases.

Squat with your feet positioned wide apart. If you can touch both hands to the floor (as shown in the photograph) then so much the better. The jump has to be committed, so make sure your head isn't directly underneath the pull-up bar; it should be set back slightly so you can lead with your chest. This is because unlike a regular strength pull-up, this is a propulsion pull-up, which means that with the right levels of commitment you'll be able to pull yourself much higher than you would normally.

1

The payback
Most people can do squats, some people can do quality pull-ups, but neither of these moves on their own will be particularly good for HIIT. By combining the two together, however, this exercise becomes extremely intense and productive, and uses every muscle in the body (including the one that's called 'ego'!).

2

3

Why this made the HIIT list

The squat is a classic exercise that can stimulate decent levels of intensity when performed with just your bodyweight. However, when you seamlessly blend it with the jump and pull phases as outlined here, it becomes a spectacular HIIT move.

"The jump has to be committed, so make sure your head isn't directly underneath the pull-up bar."

Move: Box Jumps

This explosive really works the lower body, torso, chest and shoulders. Aim for high jumps at top speed for big results.

The brief

■ Stand no more than an arm's length from your plyo box with your feet shoulder-width apart. Very quickly perform a half-squat (this is simply done to build up momentum) and swing your arms backwards.

■ As your arms reach their maximum range, swing them forwards and upwards while simultaneously springing up to jump your feet on to the plyo box.

■ You now have two options: A stand up fully or B stay in your landed position, then immediately jump back off the plyo box. Option A is more intense, and therefore preferable.

■ As soon as you make contact with the floor you should go into rebound mode by dropping into the half-squat and moving on to the next repetition.

The detail

To get intensity into box jumps you need height and speed. Jumping high requires skill, while intensity is all about going fast. To achieve the former, you need to imagine that you can jump higher than the plyo box in front of you, so look up and keep trying until you have landed on the plyo box (this sounds obvious, but trust me, visualising it makes a huge difference). If the plyo box is above knee height then you must exaggerate the

knee lift and draw them upwards until they are almost hitting your chest, spreading them slightly further than hip-width apart. In addition, to generate forward momentum you need to throw your arms up and at least to shoulder height if you are going to quickly get up on the plyo box.

When you reverse the movement from plyo box to floor it needs to be a forced action, by which I mean it's not enough to let gravity do its thing: you have to get down fast, so jump up and off the plyo box with a high level of urgency. As soon as you touch the ground, immediately move on to the next repetition.

The payback

Jumping has a profound effect upon your body. Although you are only really exerting load when your feet are in contact with the floor or the plyo box, the rapid movements of the entire body that occur while in the air are good for you. That said, it's when you are generating force that most of the gains occur, and since this exercise necessitates huge amounts of explosive strength in the lower body, torso and even your chest and shoulders, it's another win–win move!

Why this made the HIIT list

When the fitness industry adopted the plyometric boxes, however, everything was turned upside down and the movement was reversed to create this fantastic high-intensity move.

"You need to imagine that you can jump higher than the plyo box ... "

CARDIO SESSIONS ON MACHINES

Running machine: Run Mode

Maximum exertion = maximum results. Remember, running as part of a HIIT workout is no gentle jog in the park.

The brief
- Set the running machine to an incline of between 2 degrees and 8 degrees.
- Start the running machine and increase the speed to as fast as you can go. Keep your body upright and focus on moving your arms and legs quickly, using short rather than long strides.
- At the end of the interval, slow down to recover, then speed up again at the start of the next interval.

The detail
Your inspiration for 'what good looks like' should be to emulate the action and style of a good 400m runner: super-smooth, not effortless but certainly not forced either – relaxed isn't a word that you'd associate with HIIT speed running, but it's not a bad description in this case. The knee lift should be high, but there's nothing to be gained by bringing your femur (thigh bone) above horizontal since this will cause deceleration.

Your foot strike needs to complement the stride length. As a rule for sprinting on running machines, your lower leg (shin bone) shouldn't cross the imaginary vertical line that drops from your knee to the running-machine belt. If it does, you'll be clawing (pulling) at the ground rather pushing.

There should be no noticeable rotation in the torso or around the shoulders. In fact, if there is then you are most likely letting some of your power be diluted. The power you generate through your lower body can only be effectively turned into traction if your upper body is generating torsion and recoil, so pump your shoulders as fast as you can and you'll find that your legs will follow their lead.

You can't simply guess how fast is too fast, so ensure that before you actually attempt a HIIT session you do a test run to establish a range of speeds at which you feel challenged but in control.

The payback
If exercise is going to be beneficial then it needs to challenge the body, and interval running is a good way to do this. Running is a primal movement, so mechanically our bodies are equipped for it to feel instinctive. Saying that, if you don't run often I think you very quickly lose your ability to instantly switch on rhythm and control, both of which are essential to enable you to perform at the level required for HIIT,

Note: Model shown off treadmill to display correct body position so it may take non-runners some time to achieve the desired intensity.

There's no external resistance when we run, other than the force of pushing against the ground and wind resistance, so in order to get the payback of cardio-respiratory, cardiovascular and metabolic gains you need to always work at a level (speed and/or incline) that leaves you feeling 'smashed' by the end of the session.

Why this made the HIIT list

Running is awesome in every way: it's cheap, accessible, and can be done indoors or outside, by old or young, on your own or as a social activity. There are also numerous electronic devices you can use to monitor your progress, the results of which you could post on social-media sites, should you so wish. All it requires is a decent pair of trainers, fitted by a professional whenever possible, and some basic running clothing, and you're off.

"Do a test run to establish a range of speeds at which you can feel challenged but in control ... "

Running machine: Push Mode

**Attack the treadmill and work hard at maintaining
a high level of intensity for big gains.**

The brief

■ Set the running machine to an incline of between
2 degrees and 10 degrees (8 degrees seems to be
optimal). Your target body position is somewhere
between prone and upright.

■ Hold the crossbar and gradually increase the speed
until your pace is such that you can push hard against
the running belt through 90 per cent of your stride
(the last 10 per cent being when you lift your foot to
step forwards).

■ Keep you arms almost straight and focus on the
connection between the force you put through your legs
and the pressure you can feel as your hands are pushed
against the crossbar.

■ Work in this way for the duration of the interval,
then slow down to recover, before speeding up again for
the next interval.

The detail

This move isn't a mainstream activity and you'll
probably never be taught it in a gym induction. Start
slowly and build up the speed. If the running machine
seems to be dragging you along then it's moving much
too fast. Aim for a speed at which you can generate
intensity, but do so by pushing rather than as result of
what usually occurs on a running machine, which is
that you aim to keep up with the belt.

Note: Model
shown off
treadmill to
display correct
body position

On a non-motor running machine, such as the Matrix S drive, you can let your head drop below the height of your hands, but I think it's unrealistic to attempt this on a motorised machine. Instead, rather than assuming a double-handed Superman pose, keep looking up at the running machine's console since this will encourage you to keep your chest raised up just enough to keep the force on the hands and minimise the risk of falling on your face on the running belt.

The payback

I've been using the new generation of running machines that don't have a motor (self-powered running machines), the best of which have a function called 'sledge mode', which offers a phenomenal level of intensity. This is possible because the machines have such awesome braking systems that you have to generate up to 118kg (260lb) of force just to move the belt. A conventional motorised running machine isn't going to be able to deliver that kind of experience, but you can still use one to create the desired cardiac intensity and muscle fatigue.

Why this made the HIIT list

Sprinting is without doubt a skill, and it may be that you have no desire (or feel you are too big or overweight) to accomplish the level of ability required to control yourself on a fast-moving running belt. This is where 'push mode' comes into play. It's not as good as sprinting, but it's certainly no slouch. This seems like as good a time as any to roll out my favorite mantra as it sums up every exercise in the book: 'If you're moving you're improving, but if you also happen to be pushing, pulling, twisting or jumping then you are probably improving more than the guy who isn't!'

Running machine: Hike Mode

Hike your way to superb abs, glutes and hamstrings with
this demanding yet highly rewarding HIIT treadmill move.

The brief
■ Set the running machine to between 10 degrees
and the maximum incline, and start walking. Aim for
the running machine to be moving at a speed at which
you feel you have to work really hard in order to stay
close to the crossbar. If it seems you are getting on
top of the crossbar, speed up the running machine.
However, if the belt speed is pushing you towards the
back of the belt, take action to get back to the centre

"A forefoot strike is more desirable than a heel strike due to the fact that the latter causes a moment of deceleration with every step you take."

by moving more quickly. Your stride length should be such that if the running machine weren't on an incline it would look as though you were doing an exceedingly long stride, but on a 10-degree incline it should resemble a very fast walking lunge.

■ Heel strike or forefoot strike? It depends upon the speed and length of your stride, but generally a forefoot strike is more desirable than a heel one.

■ Your arms should be moving very quickly, with the power being generated in the shoulders. Hike in this way the duration of the interval, then slow down to recover, and speed up again for the next interval.

The detail

The incline is key to the success of this exercise. Fast walking on a running machine simply means keeping up with the belt whereas hiking requires you to stay ahead of the tempo or you'll come off the back of the machine. The best version of hiking looks like a very fast walking lunge rather than speed walking, with the arms moving quickly and power radiating from the chest, upper back and shoulders.

A forefoot strike is more desirable than a heel strike due to the fact that the latter causes a moment of deceleration with every step you take.

The payback

Hamstrings, glutes, abs and your upper body are all required to get you up the hill. Although the gains are admittedly more modest than those to be had in sprint or push mode, if you are able to find the balance between speed and incline you'll be demanding that your body forces itself to adapt and thus improve.

Why this made the HIIT list

It's not as good as running nor as instantly intense as pushing, but hiking is still worth doing as a way of including some HIIT in a gym regime, and may be especially appropriate on days when you are feeling fatigued or a little under the weather and wouldn't otherwise do anything.

Spin Bike: Standard Mode

Power your way through this blistering spin cycle HIIT move.
A low-impact exercise that delivers superb HITT returns.

The brief

■ Set up the bike to suit you (see p. 77). Your hand position is important rather than critical; hands should be wide and at the back of the bars when seated, in the middle or front of the handles when climbing (pushing heavy resistance) and wherever feels most comfortable when doing position-based drills or exercises.

■ Set the resistance to whatever feels comfortable for you and turn over your legs at a cadence (revolutions per minute or rpm) of around 80–90rpm while you warm up for 5 minutes. Throughout the spin, there should be a slight plantar flexion in the ankle, meaning that the ball of your foot should be lower than the heel. Never drop the heel below the pedal. The optimal drive phase (push) happens when your front foot passes over the 12 o'clock point of the cranks rotation all the way until it almost reaches the 6 o'clock point.

■ Now your legs are warm, you can start your session.

The detail

Resistance is king on a spin bike. Speed requires skill, but pushing against resistance consumes calories and is the only way to stop the flywheel from running away from you and building up pointless momentum. All the work is done by your lower body, and your ability to keep yourself balanced in the centre of the bike rather than bobbing around from side to side is the difference between looking like an amateur and a pro.

When working out how much resistance to use it is very important to first take on board that you should never have no resistance on the flywheel. If you do, the bike will run away from you. Too much resistance is preferable to none at all, in much the same way that on a running machine it is best practice to always have a slight incline rather than running on the flat. How much depends upon your fitness, but I always like to think that a spin bike is no different to my road bike, which is always experiencing friction from the road

surface and is therefore presenting me with a constant reason to have to push, even if only with a small amount of effort. At the other end of the spectrum, how much is too much? If the flywheel is so heavily loaded that you can't seamlessly move from pushing with each foot because there is a pause in the cranks rotation then it's too heavy. Reduce the resistance or stand up out of the saddle, since this will enable you to push against far greater resistance.

All three modes outlined below occur between a cadence of 60rpm and 135rpm. Any quicker is not only dangerous but also undoubtedly only achieved because the resistance is negligible.

The payback

Tabata™ was created in a lab on a bike. During its development, the athletes who participated in the trials worked so hard that they had to be caught when they finished their intervals to stop them from falling off the bikes. Riding spin bikes is therefore one of the 'easiest' ways of accessing the training zones required in HIIT, not because it is effortless (quite the contrary), but because the skill level is low compared to the level of output that even the most un-athletic individual can generate.

Why this made the HIIT list

Of all the exercises in this book, this is without doubt my favourite. It has a low impact on bones and therefore doesn't leave you feeling as if you have been beaten up the day afterwards. Spin cycles also offer the most user-friendly and controllable environments of all the HIIIT moves. Moreover, classes can be sociable, and the group dynamic and a good teacher should provide encouragement and spur you on to work harder.

"Too much resistance is preferable to none at all … "

Spin Bike: Speed Mode

This vigorous HIIT exercise is sure to fire the metabolism and is a great way to get your body into top condition.

The fastest I've ridden in a spin session is 135rpm, but in all honestly I don't like this speed for anything other than a 60-second dash; 120rpm is a far better glass ceiling for your ambitions to go fast.

To ride smoothly at speeds between 110rpm and 120rpm you need to be sitting towards the back of the seat with your hands slightly forward of your chin, arms partially bent rather than being stretched out long. It's hard to describe, but you want to create a feeling of stability within the muscles between your torso and core. I find that when I'm getting close to my top tempo I'll instinctively squeeze inwards with my inner thighs so that my knees move over my big toe rather than being above the middle of my foot.

Before you launch into a fast spin, it is important to consider how you'll slow down or stop once you reach the end of the high-intensity section. Some, but not all, spin bikes have brakes, which should be applied very gently in order to gradually slow down the flywheel – do not just slam them on except in an emergency or you could do yourself some real harm. The best way to slow down is to reduce your leg speed. This requires a little skill but is preferable to using the brake since it works your legs more through the transition phase, and besides, it may be your only option on a bike that doesn't have a brake.

Position Mode

This HIIT exercise really works the legs and gets the blood pumping. Each position works a different set of muscles.

Spin session position-based moves include lifts, jumps and hovers. To do these, ride sitting at a speed of 70–90rpm with your hands wide and towards the back or middle of the bars. Ensure that you have enough resistance to instantly slow the bike if you take your foot off the gas, then inject more pace by quickly unloading your bodyweight from the saddle (that's the jump – see photograph 2). Now maintain the additional speed, but rather than shifting you body position so you knees go in front of the pedals, keep your buttocks above the seat (photograph 2) so that you hover. If you can't feel it then you are either extending your legs too much or have shifted your body too far forwards and lost the athletic position that should be maintained during a good hover.

Resistance mode

Also known as 'beast mode', pushing a heavy resistance is the toughest of all the techniques used on a spin bike, but it's also the most productive.

Many bikes are fitted with sophisticated power meters that accurately measure how many watts are being generated by each rotation of the crank arms. It can come as something of a shock when you start riding these bikes how little power you generate when you ride with anything less than a very noticeable load on the flywheel.

Pushing a heavy load while seated is skillful and productive but can also feel as if it's slightly contrived. My suggestion is to get up off the seat and really attack the load using a combination of power and the raw aggression that seems to be easier to summon when you are standing up.

"You want to create a feeling of stability within the muscles between your torso and core ... "

Rower: Medium Resistance/Fast Strokes

Perfect a strong rowing technique and then apply some energy-sapping speed. Persevere and get strong, athletic, lean and defined.

The brief

■ Set the rower to medium resistance (number 5/6 on the airflow scale if you are on a Concept 2). Ensure that your foot straps and heel cups are comfortable and secure.

■ The first stroke should be used to get the 'boat' moving, then use the second and third stroke to accelerate to a steady HIIT rhythm of between 30–40 strokes per minute.

■ Although your arms, shoulders and upper back do play a significant role, your legs should do the majority of the work during the pull phase of the stroke. The

unloaded phases should be done quickly but not so fast that it causes your shins to go beyond vertical (if they do, not only will you lose momentum but you'll also be creating what's know as a shearing force in your knee, which is something that is best avoided).

The detail

The first three strokes will feel the heaviest; if you were a rower then you would practise these on a different resistance level, but for our purposes medium resistance (number 5/6 on the airflow scale if you are on a Concept 2) is optimal. This might surprise you,

but despite the fact that even out-of-shape people can and do row on the highest setting, I was assured by a double Olympic gold-medallist rower that the mid-range is all that is needed for fitness. In his words, the only coaching tips I really needed to adhere to were: 'Smash it with your legs and then anything that the arms add to the party is a bonus'! Can't argue with that.

Once you are flowing it will feel as if you can go really fast but don't be tempted to shorten your stroke length in pursuit of speed. Remember, the oar is only pulling you though the water when it's in the water.

The payback
You'll become fantastically resilient and able to cope with great physical demands if you get good at rowing. This is because the level of intensity (and in turn the metabolic effects) that you can achieve on a good rowing machine is phenomenal. However, you should bear in mind that rowing for a significant amount of time every week will result in the development of very broad shoulders and toned legs, but won't especially work the chest or abs. If you are hoping to achieve all-round conditioning then it is best to mix it up and just include a couple of rowing sessions for fitness' purposes.

Why this made the HIIT list
Rowing isn't a primal movement – in fact it's rather an awkward skill to learn – but if you persevere you'll have the opportunity to really challenge your body within just minutes, with the result that you'll become stronger, leaner and most of all powerful beyond all ordinary expectations.

3

"I was assured by a double Olympic gold-medallist rower that the mid-range is all that is needed for fitness."

PART FOUR

THE WORKOUTS

THE ONLY HIIT EXERCISES YOU'LL EVER NEED

HIIT is hard work, and since complexity and intensity don't sit well together, for this book I looked to evolve what I already knew and then decided where I could add an extra dimension of productivity by introducing a greater range of motion, extra planes of motion, or simply changing how the body interacts with the floor or piece of equipment.

At first glance, many of the moves found in *The HIIT Bible* might seem familiar, but the devil is in the detail … I'm not claiming that I've invented moves that are highly radical, but I'm fairly sure that many will be new to even the most experienced of exercise aficionados.

The exercise list found in this book has been designed so that it contains the only HIIT exercises you will ever need. When you get to the HIIT workouts you'll find that they consist of just four exercises, they are all fantastically efficient compound moves (meaning every repetition recruits as many muscles as possible), and they are sequenced to ensure that they complement the exercises that precede and follow after them.

Warming up

The preparation and warm-up for the session depends very much upon how you feel but you need to reach the point at which you are moving or lifting weight in a heightened sense of activeness. In order to achieve this, you should progressively work through all the major joints of the skeleton – preparing them for what is about to happen, so that by the time you start the workout you feel as if your body is working at running pace rather than walking speed.

Each session will be different for every person, but here's a suggested framework that is applicable for all of the workouts, including those later in the book that incorporate dumbbells and gym equipment.

First mobilise: Progressively move every joint from your shoulders to your ankles through as big a range as you can. Move every joint at least 10 times, or until the movement feels free-flowing, without restriction or pain.

Elevate: Generate some heat by jumping up and down (with a skipping rope if you like) or running on the spot. This can be a very productive way to prepare to exercise. Add more beats: Your heart should be beating at a high rate by the end of a warm-up session. If your heart is pumping at anything less than 100 beats per minute, you aren't invigorated enough.

Rebound: Once you feel as if your body is responding, go back through the original mobility moves but this time do them more vigorously, with speed, and so that you feel muscles rebounding at the end of movements.

Add even more beats: Generate more heat and heart beats. Run some more and jump some more. Beads of sweat may now start to form on your forehead – that's a sign that you are ready for action.

Cooling down

When you have finished a HIIT workout your heart rate will be elevated. Stopping suddenly can cause you to feel lightheaded so aim to keep moving for 1–2 minutes until your heart rate has returned to its pre-exercise pace.

The benefits of traditional end of a workout static stretches are constantly being questioned so my advice is to stretch after your HITT session to re-set your body in to its 'normal' natural posture. If you feel tight through your shoulders then stretch them, the same goes for your legs or any other area of the body. Rather than employing static stretch poses, move the targeted area through big sweeping movements until it feels as if the muscles have released.

If increasing your flexibility and quality of movement is your goal, then you actually need to treat it as stand-alone exercise session rather than tagging it on to the end of HIIT sessions. Using a foam roller is another tactic for maintaining mobility and working on the flexibility and your range of motion.

BODYWEIGHT MOVES

Yoga, Pilates, Zumba®, aerobics, INSANITY®, Hard Corps ... the list of possible exercise choices goes on. But what do all these workouts have in common? Answer: they are all based predominantly around bodyweight moves. But isn't nearly every human movement a bodyweight exercise? Well, yes, but they don't offer scalable benefits. Walking, for example, is a bodyweight move, but over thousands of years we have developed such an efficient way of doing it that we need to walk at a vigorous rate for it to become a valid form of intensity exercise.

In my opinion, for a bodyweight move to be classed as an exercise it has to challenge you. It should engage multiple muscles while working across multiple joints and, wherever possible, demand that the body fires or functions through all three planes of motion at the same time.

There are hundreds of bodyweight moves to choose from, but this book features a selection that offer cardio, general-strength and core-strength benefits. Some of these are very dynamic and athletic, so look around you before you start and check there is enough space: a 2 x 2m (6.5 x 6.5ft) area should suffice. Also, check the flooring surface you are exercising on is safe and remember that carpets, rugs and floor tiles have rarely been designed for high-intensity workouts.

Each of the bodyweight workouts will take up about 30 minutes of your day. This breaks down as 5 minutes to warm up, 5 minutes to cool down, and the rest is either flat-out workout or recovery/active recovery. The workouts have intentionally been kept to as few exercises as possible to ensure that productivity is high, so you can maintain your momentum by going into autopilot on a second set rather than thinking about what exercise comes next.

Intensity modes

Each of the bodyweight workout programmes consists of four HIITastic exercises and can be attempted in three different modes, as outlined in detail below:

Mode one: you are super-motivated

This is going to be tough. You should be looking to deliver As Many Reps As Possible (AMRAP) for 30 seconds for each move, followed by 30 seconds' total recovery (total recovery means you make no movements at all, and do nothing but catch your breath). Repeat everything twice (2 sets of everything).

SET 1
Exercise 1 AMRAP for 30 seconds, total recovery for 30 seconds
Exercise 2 AMRAP for 30 seconds, total recovery for 30 seconds
Exercise 3 AMRAP for 30 seconds, total recovery for 30 seconds
Exercise 4 AMRAP for 30 seconds, total recovery for 30 seconds

SET 2

Exercise 1 AMRAP for 30 seconds, total recovery for 30 seconds
Exercise 2 AMRAP for 30 seconds, total recovery for 30 seconds
Exercise 3 AMRAP for 30 seconds, total recovery for 30 seconds
Exercise 4 AMRAP for 30 seconds, total recovery for 30 seconds
Cool down and finish

Mode two: you want to get this over with quickly

HIIT isn't fun, and there is no shame in wanting to get it over and done with. For
this mode, each move needs to be performed AMRAP for 20 seconds, followed by 10
seconds' total recovery. Repeat everything twice (2 sets of everything).

SET 1

Exercise 1 AMRAP for 20 seconds, total recovery for 10 seconds
Exercise 2 AMRAP for 20 seconds, total recovery for 10 seconds
Exercise 3 AMRAP for 20 seconds, total recovery for 10 seconds
Exercise 4 AMRAP for 20 seconds, total recovery for 10 seconds

SET 2

Exercise 1 AMRAP for 20 seconds, total recovery for 10 seconds
Exercise 2 AMRAP for 20 seconds, total recovery for 10 seconds
Exercise 3 AMRAP for 20 seconds, total recovery for 10 seconds
Exercise 4 AMRAP for 20 seconds, total recovery for 10 seconds
Cool down and finish

Mode three: you want to work hard but more slowly, and for longer

This mode allows you to work out at a slower pace with increased recovery times. This doesn't mean you are taking it easy though – you're just resting for longer so that you can put more into the 30-second work periods. For this mode, each move needs to be performed AMRAP for 30 seconds, followed by 90 seconds' active recovery (active recovery means you keep moving). Repeat everything twice (2 sets of everything).

SET 1

Exercise 1 AMRAP for 30 seconds, active recovery for 90 seconds
Exercise 2 AMRAP for 30 seconds, active recovery for 90 seconds
Exercise 3 AMRAP for 30 seconds, active recovery for 90 seconds
Exercise 4 AMRAP for 30 seconds, active recovery for 90 seconds

SET 2

Exercise 1 AMRAP for 30 seconds, active recovery for 90 seconds
Exercise 2 AMRAP for 30 seconds, active recovery for 90 seconds
Exercise 3 AMRAP for 30 seconds, active recovery for 90 seconds
Exercise 4 AMRAP for 30 seconds, active recovery for 90 seconds
Cool down and finish

Bodyweight moves (no equipment): HIIT workout programmes

Bodyweight moves (no equipment):
HIIT workout 1

This vigorous HIIT workout features a range of exercises designed to work a range of muscles, get your heart racing and make your body feel fantastically energised. Choose your mode, make a note of the intensity and intervals, then get going with the exercises.

THE EXERCISES
Run-ups (see pp. 102–103)
Thrusters (see pp. 98–99)
Big Jump Jacks (see pp. 106–107)
Rabbit (see pp. 108–109)

Bodyweight moves (no equipment): HIIT workout 2

Each of these four HIIT exercises will give you an almighty workout. Remember though, the harder you work, the better the results, and the more great you will feel! Choose your mode, make a note of the intensity and intervals, then get going with the exercises.

THE EXERCISES
Plunges (see pp. 104–105)
Gecko (see pp. 100–101)
Skater (see pp. 116–117)
Bangbang Firing (see pp. 110–111)

Bodyweight moves (no equipment): HIIT workout 3

This workout features another four awesome exercises. As with all HIIT moves, it is important to execute them accurately, otherwise you won't gain all the results your efforts deserve. Choose your mode, make a note of the intensity and intervals, then get going with the exercises.

THE EXERCISES
Big Jump Jacks (see pp. 106–107)
Rabbit (see pp. 108–109)
Run-ups (see pp. 102–103)
Thrusters (see pp. 98–99)

Bodyweight moves (no equipment): HIIT workout 4
Tackling a range of HIIT exercises over time means your body gets a complete workout. It also means you'll never get bored, since different workouts offer different physical challenges. Choose your mode, make a note of the intensity and intervals, then get going with the exercises.

THE EXERCISES
Bangbang Firing (see pp. 110–111)
Arrows (see pp. 114–115)
Skater (see pp. 116–117)
Plunges (see pp. 104–105)

Bodyweight moves (no equipment): HIIT workout 5
This workout is made up of four fantastic HIIT moves. Remember, the quicker you work and the more control you exert, the faster you will see results. Choose your mode, make a note of the intensity and intervals, then get going with the exercises.

THE EXERCISES
BurpEE (see pp.112–113)
Gecko (see pp. 100–101)
Big Jump Jacks (see pp. 106–107)
Rabbit (see pp. 108–109)

Bodyweight moves intensified with dumbbells HIIT

For this section, a number of fantastic HIIT moves have been enhanced by the addition of a single set of dumbbells for a more vigorous workout and optimised results. The weights used don't have to be huge to be challenging and effective. I suggest that 5kg (11lb), 7.5kg (16½lb) or occasionally 10kg (22lb) dumbbells are used for all of the moves. We are all different, though, so some people may need heavier weights, while a good many will actually find that 5kg (11lb) is going to feel heavy on some of the moves, especially when arms are fully extended.

It is important to ensure your exercise area is big enough to accommodate both you and the weights. You should also consider the floor surface. An accidentally dropped dumbbell could dent or damage one that is wooden or made from a delicate or valuable material, so it is best not to work in these conditions.

Most dumbbell exercises you see in the gym are very basic and focus on isolating muscles and increasing the weight until the movement is going at a slow pace. All of the exercises shown in this section are the polar opposite to those – instead, these are HIIT bodyweight moves intensified with dumbbells that encourage muscles located throughout the body to fire, especially the deep core muscles.

Many of you may be wondering why I suggest using dumbbells rather than kettlebells for HIIT. The answer is that while kettlebells can be great for HIIT, you need to become very proficient in their use before you can integrate them into a session – learn it, then work it! Dumbbells are simpler to use, providing a great and highly productive introduction to HIIT with weights. Working with kettlebells can be a seen as the next challenge.

Bodyweight moves intensified with dumbbells: HIIT workout 1

Always ensure you are working in enough space to accommodate your body movements and the swinging of the dumbbells. These moves are going to work you hard and the last thing you want to worry about is causing damage to yourself or the workout area. Choose your mode, make a note of the intensity and intervals, then get going with the exercises.

THE EXERCISES
Bottom Up (see pp. 120–121)
Halo (see pp. 138–139)
Bell Ringer (see pp. 126–127)
Stones Lift (see pp. 130–131)

Bodyweight moves intensified with dumbbells: HIIT workout 2

As with all HIIT moves, if you execute these properly at a high level of intensity, you'll burn more fat in less time. Choose your mode, make a note of the intensity and intervals, then get going with the exercises.

THE EXERCISES
Double Thruster (see pp. 124–125)
Swing and Step (see pp. 128–129)
V-jumps (see pp. 132–133)
Turk Thruster (see pp. 136–137)

Bodyweight moves intensified with dumbbells: HIIT workout 3

This selection of exercises delivers a truly astonishing workout. Remember, controlled execution delivered at an intense speed is the key to fantastic results from HIIT. Choose your mode, make a note of the intensity and intervals, then get going with the exercises.

THE EXERCISES
Bell Ringer (see pp.126–127)
Shot-Put Driver (see pp. 134–135)
Halo (see pp. 138–139)
Split and Rip (see pp. 122–123)

Bodyweight moves intensified with dumbbells: HIIT workout 4

The joy of HIIT workouts is that they are fast, flexible and fun. Really push yourself when executing the exercises and abide by the recovery times. Choose your mode, make a note of the intensity and intervals, then get going with the exercises.

THE EXERCISES

V-jumps (see pp. 132–133)
Turk Thruster (see pp. 136–137)
Bottom Up (see pp. 120–121)
Swing and Step (see pp. 128–129)

Bodyweight moves intensified with dumbbells: HIIT workout 5

Even the most dedicated athlete can find some fitness sessions tough going. Just work through them and concentrate on how fantastic you will feel when the HIIT workout is over – and how good you will look after a few months of doing HIIT. Choose your mode, make a note of the intensity and intervals, then get going with the exercises.

THE EXERCISES

Split and Rip (see pp. 122–123)
Stones Lift (see pp. 130–131)
Double Thruster (see pp.124–125)
Shot-put Driver (see pp. 134–135)

GYM-BASED MOVES

In a health club or gym, the chances are that the free weights are in one place, all the cardio is in a zone of its own, and the fun stuff such as TRX and plyo boxes are in a space called something like the Functional Zone or Circuit Space. All of the gym-based workouts except for workout 5 start on a piece of cardiovascular equipment, where you have your longest interval, followed by two more exercises performed elsewhere in the gym. Gym-Based HIIT Workout 5 is a 'dry triathlon', during which you row, ride and run.

Gym-based HIIT workout modes

The workout modes for gym-based HIIT exercises differ from those used for the bodyweight and bodyweight intensified with dumbbells programmes. These modes have been designed to allow enough time for recovery and for moving from equipment station to equipment station.

Mode 1: four sets of everything

Even if you are huffing and puffing mid-HIIT workout (as you should be), then the chances are that you will find this workout mode easy to remember: four sets of everything.

Exercise 1 for 1 minutes, recover/relocate for 1 minute

Exercise 2 for 1 minute, recover/relocate for 1 minute

Exercise 3 for 1 minutes, recover/relocate for 1 minute

Return to Exercise 1 and repeat until you have performed four sets (done four rotations of everything)

Cool down and finish

Mode 2 – repeat sets

As the title suggests, this mode requires you to repeat each exercise. Recovery between each HIIT exercise lasts for exactly half the time spent exercising.

Exercise 1 for 2 minutes, recover/relocate for 1 minute

Exercise 1 for 2 minutes, recover/relocate for 1 minute

Exercise 2 for 1 minute, recover/relocate for 30 seconds

Exercise 2 for 1 minute, recover/relocate for 30 seconds

Exercise 3 for 1 minute, recover/relocate for 30 seconds

Exercise 3 for 1 minute, recover/relocate for 30 seconds

Cool down and finish

Mode 3: 3 x 3 power sets

For this workout programme you repeat each exercise three times for a varying length of time.

Exercise 1 for 3 minutes, recover/relocate for 90 seconds

Exercise 1 for 2 minutes, recover/relocate for 60 seconds

Exercise 1 for 1 minute, recover/relocate for 60 seconds

Exercise 2 for 90 seconds, recover/relocate for 90 seconds

Exercise 2 for 60 seconds, recover/relocate for 60 seconds

Exercise 2 for 30 seconds, recover/relocate for 30 seconds

Exercise 3 for 90 seconds, recover/relocate for 90 seconds

Exercise 3 for 60 seconds, recover/relocate for 60 seconds

Exercise 3 for 30 seconds, recover/relocate for 30 seconds

Cool down and finish

Gym-based HIIT workout programmes

Gym-based HIIT: workout 1

This workout consists of three brilliantly intense exercises, with each one majoring on different muscles. Practise your technique in advance of starting the programme to ensure you get the best results from all of your efforts. Choose your mode (four sets of everything, repeat sets or 3 x 3 power sets), make a note of the intensity and intervals, then get going with the exercises.

THE EXERCISES

Running Machine (see pp. 154–159)
Squat Rack Drop and Row Combo (see pp. 142–143)
Hang Tough Combo (see pp. 148–149)

Gym-based HIIT: workout 2

This workout utilises the spin bike, TRX system and plyo box for a HIITastic workout. Remember, the time spent travelling between and setting up the equipment is your recovery time, and it is important to do this within the stipulated recovery time. Choose your mode (four sets of everything, repeat sets or 3 x 3 power sets), make a note of the intensity and intervals, then get going with the exercises.

THE EXERCISES

Spin Bike – hover (see pp. 160–163)
TRX Pike Pull and Y-fly (see pp. 144–145)
Box Jumps (see pp. 150–151)

Position mode on the spin bike

For this spin-bike exercise I want you to focus on riding position 'jumps' that immediately blend into a hover. Ride sitting at a speed between 70–90RPM, with hands wide and towards the back or middle of the bars. It is important to ensure that you have just the right resistance so that you can instantly slow down the bike and then inject more pace by quickly unloading your bodyweight from the saddle (that's the jump), then maintain the additional speed. Rather than shifting your body position so your knees go in front of the pedals, keep your butt above the seat. When I'm teaching this in a group I describe the position as being one in which 'you feel the beak of the saddle kissing you on the butt'. If you can't feel the beak then you are either extending your legs too much or have shifted your body too far forwards and have lost the athletic position that is a good hover.

Gym-based HIIT: workout 3

As with all the workouts in this book, take your time to carefully learn each component of each exercise used before you start. This will mean you can do a harder workout that achieves far better results. Choose your mode (four sets of everything, repeat sets or 3 x 3 power sets), make a note of the intensity and intervals, then get going with the exercises.

THE EXERCISES
Rower – medium resistance/fast strokes (see pp. 164–165)
Hang Tough Combo (see pp. 148–149)
Squat Rack Drop and Row Combo (see pp. 142–143)

Gym-based HIIT: workout 4

HIIT is a highly demanding form of exercise and while it can be great fun, we all have tough moments. Work through these and once the workout is over any displeasure will be forgotten. Choose your mode (four sets of everything, repeat sets or 3 x 3 power sets), make a note of the intensity and intervals, then get going with the exercises.

THE EXERCISES

Running machine – push mode (see pp. 156–157)
TRX Skydive (Sumo Squat – Shoulder Press Combo) (see pp. 146–147)
Box Jumps (see pp. 150–151)

Gym-based HIIT: workout 5

The format of this dry triathlon-themed workout differs from the others, since it is based on you achieving AMRAP. That said, don't be a hero/fool: stop before you fall over! The format is a downwards spiral of time during the work but a set amount of recovery.

THE EXERCISES

Rower – medium resistance/fast strokes AMRAP for 4 minutes (see pp. 164–165); recover/relocate for 60 seconds
Spin Bike – speed mode AMRAP for 3 minutes (see pp. 162–163); recover/relocate for 60 seconds
Running Machine – run mode AMRAP for 2 minutes (see pp. 154–155); recover/relocate for 60 seconds
Repeat AMRAP for a maximum of three rounds

"The list of gains is extensive and these can be viewed as the building blocks for long- and short-term improvements to our body function."

OUTDOOR MOVES

HIIT the ground running

It sounds obvious, but if you want to run fast, you need to run fast. Most runners do traditional speed work, for which the aim is to achieve near race pace over distances of 400m or more, with recovery periods equal to the length of the burst (out on the road, lamp posts come in handy since you can use them as markers that define the length of your burst, then slowly jog back to the first one as you recover).

In practice, the view is that speed is invaluable but not when it is to the detriment of flow (long-distance runners tend to flow, while sprinters force). 'For the athlete who's already doing intervals,' says Martin Gibala, PhD, a professor of kinesiology at McMaster University, 'upping the intensity with short bursts of speed may provide new benefits.' This is because your cardiovascular system gets stronger and pushes more oxygen-rich blood through your body, and muscles get better at using that oxygenated blood. Your stride also becomes more efficient as coordination between the muscles and nervous system improves. The perks may even extend to reducing your risk for chronic diseases by improving blood-sugar control.

Sprinting, if you aren't used to it, does increase the risk of injury, however. 'You need to have developed both strength and flexibility and have a solid base of both mileage and speed work to safely do this training', says Joe McConkey, MS, an exercise physiologist and coach at the iconic Boston Running Center.

'You're ready for HIIT workouts if you've been running four to five times a week for at least four months, regularly doing some runs at paces 60 to 90 seconds per mile faster than easy pace, and completing a weekly long run of at least 50 minutes. In terms of strength and flexibility, you should be able to hold a squat position for 90 seconds and, while standing, grab and touch your heel to your butt, feeling only a minor stretch in your quad. Start with one HIIT session a week, and build up to no more than two in a 10-day period.'

I totally agree with Joe's strategy and always err on the side of caution when introducing HIIT to a distance-runner, because in my experience long and slow runners are more prone to niggling injuries, whereas sprinters are usually of a build that is more boom or bust – in other words, they are harder to break but when they do it's more 'spectacular'.

Running on the track

High-intensity track sessions move the muscles through the full range of motion, particularly in the hips, improving elasticity and enhancing coordination between your nervous system and muscles. Over time, you'll develop a more efficient stride at all tempos. To really optimise this process, middle-distance and high-mileage runners should take a page out of the sprinters' textbook and rather than just running miles they should practise drills, which are broken-down components of the complete running action. There's a huge range of drills, but the classics are knee drives, heel flicks, long-stride bounding and short-stride high-speed footwork. If you are time poor, use HIIT sessions during the week and enjoy your longer runs over the weekend.

TRACK HIIT SESSION

A 400m running track is perfectly organised and standardised so you can turn up, warm up, do your session and be on your way home within 45 minutes of arriving at the track.

Warm up with some mobility exercises and at least one lap of jogging.
Start at the 250m mark on the back straight and begin your Track HIIT session by sprinting from the 250m line to the finish line.
For the recovery section, continue around the track at a slow walking pace until you get back to the 250m mark.

You can allow yourself between three and four times as much recovery time as the sprint section took, e.g. If the sprint took 40 seconds, take at least 120 seconds and up to 160 seconds recovery/walk back time before starting the next sprint. Aim to complete eight sets of sprint/recovery run/walks.

Running off-road
One man's fun off-road trail is another man's muddy path that gets his new shoes dirty. Personally, I don't think off-road trails are the right environment in which to be doing HIIT sessions, but you could argue that running fast over softer, less-groomed terrain such as forest paths, beach trails or grass can increase agility and athleticism. For me, however, it's probably a risk that not worth taking – it's all too easy to twist an ankle or take a tumble, meaning you have to focus more on the terrain than the actual exercise. The exception to this is the beach; running in bare feet down near the water where the sand is soft but not impossible to gain traction on can provide a thrilling HIIT experience.

OFF-ROAD HIIT SESSION

Choose a location that is either a flat circuit or one that includes a moderate hill. If you want to go from A to B rather than do laps then you'll have to accept that every interval will be different because of the varying terrain, inclines and declines. A loop that is between 400 and 800 metres is ideal.

Warm up with some mobility exercises and at least one lap of jogging. Using either a stopwatch to time the interval or landmarks such as trees or lamp posts, sprint for 30 seconds then slow down gradually and walk for 90 seconds. Aim to repeat this 8 times.

You can of course sprint for longer if you wish, however if you are running on unmade paths the longer you sprint for the more fatigued you'll become, which increases the risk of losing your footing, so focus on quality as well as quantity – but above all keep yourself safe.

Running up inclines

A pavement/sidewalk with a serious incline is a great venue for super-fast tempo leg speeds and speed work. This is because, compared with a flat surface, hills alter the impact on your legs and limit your range of motion, thereby in theory lowering the risk of muscle pulls and strains. As always, proceed with caution.

HILL HIIT SESSION

1 Run up your chosen hill at a moderate pace for 30 seconds. Walk down the hill for your recovery. Repeat twice more, so you do three efforts in total.
2 When this feels controlled and comfortable, progress to 4 x 1 minute near all-out efforts with a downhill jog and an additional 30–60 seconds' jogging or walking rest.
3 Over time, add additional reps and/or extend the effort time up to 2 minutes. When you have mastered this, go and find a steeper hill.

When you are running outside there are inherent dangers. For instance, paths that appear perfectly flat when you are walking on it can suddenly become far less predictable underfoot when you are doing a HIIT sprint session. Be selective about where you run when doing HIIT, not just about the terrain but also be aware of pedestrians and drivers and cyclists. It is essential that you stay in control at all times.

HIIT Cycling

If you are a dedicated cyclist like me, then the chances are you are already doing HIIT, even if this is just an unintentional consequence of your riding style, influenced by the people you ride with, or your constant pursuit of self-preservation on the roads (going fast and slow to avoid being knocked off your bike!)

All the cycling-related research carried out (on the actual road rather than on a turbo trainer) shows that the benefits of HIIT are at least as good as those of base training (riding with purpose on a variety of flats, inclines/declines), and that they can be achieved with far less time on the bike. This, therefore, makes HIIT a tempting approach for anyone who has limited time – which, in reality could mean most of us. One important consideration, though, is that it won't always be possible to recreate the levels of intensity that you can achieve on a turbo trainer or spin bike at the gym once you are out on the road. This is because roads are dangerous, and you should proceed with caution. Even professional riders have horrible crashes on training rides and they often have teammates and support staff around them to minimise the risks – you most likely don't. So, stay aware of what's going on around you and please don't do a Tour de France-style stage sprint finish weaving all over the place.

Essentially, a HIIT session on your road bike is an interval session for which the intervals are undertaken above threshold (your comfort zone). In fact, you may even be doing HIIT without knowing it by sprinting between successive sets of traffic lights or attacking climbs on your favourite loop. You don't really need to decide between HIIT and base training; simply combine the characteristics of HIIT within a base-training ride. If you are time poor, use HIIT sessions during the week and enjoy your longer endurance rides over the weekend.

My HIIT road sessions total an hour but obviously I'm not fully exerting myself for that entire time. A typical session consists of a 20-minute initial cruise then, once fully warmed-up, I am generally in the mood for something more intense. The HIIT session's sprint efforts are very short and you should aim for power or heart-rate targets. The following workout is designed for you to ride as hard as you can for the predetermined time:

HIIT cycling workout

Warm up for 20 minutes

30 seconds' seated maximum effort, low/medium cadence with as heavy a gear as you can manage

30 seconds' rest

30 seconds' seated maximum effort, low/medium cadence with as heavy a gear as you can manage

30 seconds' rest

30 seconds' seated maximum effort, low/medium cadence with as heavy a gear as you can manage

30 seconds' rest

Recover by cruising for 2–4 minutes at an easy tempo

Repeat the entire high-intensity intervals and rest period three or four times

Cool down and finish

The variables and options are endless, but if you're looking to add an additional HIIT session into your weekly schedule, then the length of the intervals can be adjusted to up to 3 minutes. Remember, though, that one of the 'I's in HIIT stands for Interval, so ensure that this doesn't become a fast-tempo session with no peaks and troughs. This means that as the peaks become longer, you should reward yourself with more recovery. The easiest way to time-manage this is to work to a ratio of 1:1 – for instance, do a 3-minute AMRAP burst interval followed by a 3-minute recovery period.

PART FIVE

HIIT ASSESSMENT & TRAINING CHALLENGES

HIIT ASSESSMENT

'If you aren't assessing, then you are just guessing.' These are the wise words of renowned holistic therapist Paul Chek, and this is a mantra that can be applied to any HIIT workout programme.

Throughout my years as a track athlete, I would always know my exact Personal Best (PB) for various track distances. This is because a PB is memorable due to it being a measurement of all the hard work you put in to the preceding 52 weeks of the year. A PB gives you a target to beat and is a great means of assessing your fitness development. In short, if you can keep bettering your PB, you can be assured that you are getting fitter.

The HIIT Test

The HIIT Test is a simple way of setting your own PB and once you have you will be able to accurately assess your HIIT progress. Whatever your first HIIT Test score, if it gets better every time you retest yourself then you are doing better than the 80 per cent of the population who do little or nothing to look after their body. As a guide, unless you have started this body project as an already extremely fit individual then it's not all about your initial HIIT Test – it's how much you improve from test to test. Any improvement is an achievement but an increase of 5 per cent or more is awesome.

The HIIT Test should be an 'event' and is best conducted when you have a quality 30 minutes away from any distractions. The HIIT Test is simple, it's five exercises from the bodyweight moves that feature in the book. You do each exercise for exactly one minute and every repetition that you do counts as one point. You then get two minutes' recovery between each exercise, which should be enough to either write down or note your score on your phone. At the end, add up all the five scores and that number total is your HIIT Test score.

FastLap®

There is also a FastLap® test. This is optional and entails you needing to plot out
a running route that will take you no more than three minutes to run around. The
route could be a section of your local park or a lap of the streets outside your home
or office.

You should look at running your FastLap® route as fast as you can, and once
competed you should note your time. This is your FastLap® score.

Taking the HIIT Test®

First you MUST warm up!

Mobilise your entire body, working down from the neck, shoulders, upper back,
mid spine, pelvis, hips, and knees to the ankles. Now run on the spot, gradually at
first, and then more assertively until you feel both hot and also have increased your
heart rate to an exercise tempo.

Do each of the challenge exercises for one minute, going as fast as you can but
without sacrificing good technique. Recover for two minutes before completing the
next challenge, keep a note of each score then add them together when you finish.

The HIIT Test ®

TASK ONE	TASK TWO	TASK THREE	TASK FOUR	TASK FIVE
Thrusters for 1 minute = score	Gecko for 1 minute = score	Plunges for 1 minute = score	Rabbit for 1 minute = score	Arrows for 1 minute = score
(see pp. 98–99)	*(see pp. 100–101)*	*(see pp. 104–105)*	*(see pp. 108–109)*	*(see pp. 114–115)*
Recover for 2 minutes	Recover for 2 minutes	Recover for 2 minutes	Recover for 2 minutes	Recover for 2 minutes

Total HIIT Test® score

Optional FastLap® time

HIIT TRAINING CHALLENGE

So how often should you work out in a week? Once, twice, six times, or even seven? Everybody is different, I have formulated these two HIIT 14- and 28- Day HIIT Training Challenges to help you on your HIIT exercise journey.

In my time as a personal trainer, I have always found that clients responded much better to tasks that carried obligations. So as part of your workout commitment you might like to put aside an amount of money each day on the basis that if you stick to the daily HIIT challenge you'll get your stake back, but if you fail to follow the plan then you have to give the money to someone or something that actually doesn't deserve it (suggesting it goes to charity would be nice but it would ruin the sense of jeopardy, so it has to be comparable to throwing the money down the drain).

Once you have committed to the challenge, look ahead in your schedule to ensure that you have a straight 14 or 28 days available, set yourself a time slot each day and then just do it. The challenge for each day is very simple: you do each exercise on the schedule for a set amount of time, you then move on to the next exercise on the list. You should then continue to work through all of the exercises until the overall workout time is up.

Day One (excluding the warm up) requires you to exercise for 15 minutes.

Example:

DAY ONE

Thrusters for 20 seconds *(see pp. 98–99)* Recover for 40 seconds	Gecko for 20 seconds *(see pp. 100–101)* Recover for 40 seconds	Run-ups for 20 seconds *(see pp. 102–103)* Recover for 40 seconds	BurPEE for 20 seconds *(see pp. 112–113)* Recover for 40 seconds

Repeat AMRAP for the time remaining of the 15 minutes.

It's that simple. It is important to remember to warm up before starting the challenge each day and you will see there are recovery days built in to the challenge. These recovery days are non-negotiable so don't bank, save or switch them to alternative days. Both the 14- and 28- Day challenges are exactly that: 'a challenge', so approach them with caution and never battle on through illness or injury.

If you are you are new to exercise, then the level of activity performed during the challenges will mean you'll be exploring a fine line between feeling great and being overwhelmed, so listen to your body and always let common sense prevail. Most people will go through a number of emotions when they make a change to their lifestyle, but unlike dramatic lifestyle changes where you are giving something up, for example, smoking, you won't be fighting against any negative chemically induced challenges. An exercise challenge, while tough, is generally a win-win situation, but knowing that most people go through a range of emotions when starting something new might just help you overcome any temptation to give up.

14-DAY
'ZERO TO HERO' CHALLENGE

The 14-Day 'Zero to Hero' Challenge is designed to make you feel better and realise that exercise doesn't have to eat in to huge sections of your day.

DAY ONE

On day one of the 14-Day 'Zero to Hero' challenge, you should take the HIIT Test. You will take the HIIT Test again at the end of the challenge so you can measure your progress. Don't forget to take your 'before' selfie!

The HIIT Test
(see p. 204)

DAY TWO

Thrusters for 20 seconds	Gecko for 20 seconds	Run-ups for 20 seconds	BurPEE for 20 seconds
(see pp. 98–99)	*(see pp. 100–101)*	*(see pp. 102–103)*	*(see pp. 112–113)*
Recover for 40 seconds	Recover for 40 seconds	Recover for 40 seconds	Recover for 40 seconds

Repeat AMRAP for the time remaining of the 15 minutes.

DAY THREE

Thrusters for
20 seconds
(see pp. 98–99)
Recover for
40 seconds

Gecko for
20 seconds
(see pp. 100–101)
Recover for
40 seconds

Run-ups for
20 seconds
(see pp. 102–103)
Recover for
40 seconds

BurPEE for
20 seconds
(see pp. 112–113)
Recover for
40 seconds

Repeat AMRAP for the time remaining of the 15 minutes.

DAY FOUR

Thrusters for
20 seconds
(see pp. 98–99)
Recover for
40 seconds

Gecko for
20 seconds
(see pp. 100–101)
Recover for
40 seconds

Run-ups for
20 seconds
(see pp. 102–103)
Recover for
40 seconds

BurPEE for
20 seconds
(see pp. 112–113)
Recover for
40 seconds

Repeat AMRAP for the time remaining of the 15 minutes.

DAY FIVE

Rest day from HIIT but not from all activity. Stretch, do mobility exercises and be as active as you can without actually 'working out'.

DAY SIX

Plunges for 20 seconds (see pp. 104–105) Recover for 40 seconds	Big Jump Jacks for 20 seconds (see pp. 106–107) Recover for 40 seconds	Run-ups for 20 seconds (see pp. 102–103) Recover for 40 seconds	Bangbang Firing for 20 seconds (see pp. 110–111) Recover for 40 seconds	Arrows for 20 seconds (see pp. 114–115) Recover for 40 seconds

Repeat AMRAP for the time remaining of the 20 minutes.

DAY SEVEN

Plunges for 20 seconds (see pp. 104–105) Recover for 40 seconds	Big Jump Jacks for 20 seconds (see pp. 106–107) Recover for 40 seconds	Run-ups for 20 seconds (see pp. 102–103) Recover for 40 seconds	Bangbang Firing for 20 seconds (see pp. 110–111) Recover for 40 seconds	Arrows for 20 seconds (see pp. 114-115) Recover for 40 seconds

Repeat AMRAP for the time remaining of the 20 minutes.

DAY EIGHT

Plunges for 20 seconds (see pp. 104–105) Recover for 40 seconds	Big Jump Jacks for 20 seconds (see pp. 106–107) Recover for 40 seconds	Run-ups for 20 seconds (see pp. 102–103) Recover for 40 seconds	Bangbang Firing for 20 seconds (see pp. 110–111) Recover for 40 seconds	Arrows for 20 seconds (see pp. 114–115) Recover for 40 seconds

Repeat AMRAP for the time remaining of the 20 minutes.

DAY NINE

Rest day from HIIT but not from all activity. Stretch, do mobility exercises and be as active as you can without actually 'working out'.

DAY TEN

Run-ups for	Gecko for	Skater for	Rabbit for
30 seconds	30 seconds	30 seconds	30 seconds
(see pp. 102–103)	*(see pp. 100–101)*	*(see pp. 116–117)*	*(see pp. 108–109)*
Recover for	Recover for	Recover for	Recover for
60 seconds	60 seconds	60 seconds	40 seconds

Repeat AMRAP for the time remaining of the 20 minutes.

DAY ELEVEN

Thrusters for	Gecko for	Run-ups for	BurPEE for
20 seconds	20 seconds	20 seconds	20 seconds
(see pp. 98–99)	(see pp. 100–101)	(see pp. 102–103)	(see pp. 112–113)
Recover for	Recover for	Recover for	Recover for
40 seconds	40 seconds	40 seconds	40 seconds

Repeat AMRAP for the time remaining of the 15 minutes.

DAY TWELVE

Thrusters for	Gecko for	Run-ups for	BurPEE for
20 seconds	20 seconds	20 seconds	20 seconds
(see pp. 98–99)	(see pp. 100–101)	(see pp. 102–103)	(see pp. 112–113)
Recover for	Recover for	Recover for	Recover for
40 seconds	40 seconds	40 seconds	40 seconds

Repeat AMRAP for the time remaining of the 15 minutes.

DAY THIRTEEN

Thrusters for	Gecko for	Run-ups for	BurPEE for
20 seconds	20 seconds	20 seconds	20 seconds
(see pp. 98–99)	(see pp. 100–101)	(see pp. 102–103)	(see pp. 112–113)
Recover for	Recover for	Recover for	Recover for
40 seconds	40 seconds	40 seconds	40 seconds

Repeat AMRAP for the time remaining of the 15 minutes.

DAY FOURTEEN

Do the HIIT Test
Compare your score to the start of the challenge, any improvement is good in such a short period of time, however anything greater than a 5–10% improvement is awesome. Take an 'after' selfie and just see how good you look compared to the 'before' selfies taken 14 days ago.

Once you have completed the 14-Day 'Zero to Hero' Challenge I suggest you either plan to use the other programmes in *The HIIT Bible*, especially if you have access to the relevant fitness equipment as this will increase your exercise options and potentially accelerate your gains. Alternatively, dive in head first to the 28 day Zero to Super Hero Challenge. Again, don't forget to take your 'before' selfie!

28-DAY 'ZERO TO SUPER HERO' CHALLENGE

Start the 28-Day 'Zero to Super Hero' Challenge by taking the HIIT Test. You will take the HIIT Test again at the end of the challenge so you can measure your progress. Don't forget to take your 'before' selfie!

DAY ONE

The HIIT Test *(see p. 204)*

DAY TWO

Run-ups for	Plunges for	Rabbit for	Skater for
20 seconds	20 seconds	20 seconds	20 seconds
(see pp. 102–103)	*(see pp. 104–105)*	*(see pp. 108–109)*	*(see pp. 116–117)*
Recover for	Recover for	Recover for	Recover for
40 seconds	40 seconds	40 seconds	40 seconds

Repeat AMRAP for the time remaining of the 15 minutes.

DAY THREE

Run-ups for	Plunges for	Rabbit for	Skater for
20 seconds	20 seconds	20 seconds	20 seconds
(see pp. 102–103)	*(see pp. 104–105)*	*(see pp. 108–109)*	*(see pp. 116–117)*
Recover for	Recover for	Recover for	Recover for
40 seconds	40 seconds	40 seconds	40 seconds

Repeat AMRAP for the time remaining of the 15 minutes.

DAY FOUR

Run-ups for	Plunges for	Rabbit for	Skater for
20 seconds	20 seconds	20 seconds	20 seconds
(see pp. 102–103)	*(see pp. 104–105)*	*(see pp. 108–109)*	*(see pp. 116–117)*
Recover for	Recover for	Recover for	Recover for
40 seconds	40 seconds	40 seconds	40 seconds

Repeat AMRAP for the time remaining of the 15 minutes.

DAY FIVE

Rest day from HIIT but not from all activity. Stretch, do mobility exercises and be as active as you can without actually 'working out'.

DAY SIX

Big Jump Jacks for 20 seconds (see pp. 106–107) Recover for 40 seconds	Thrusters for 20 seconds (see pp. 98–99) Recover for 40 seconds	Gecko for 20 seconds (see pp. 100–101) Recover for 40 seconds	Bangbang Firing for 20 seconds (see pp. 110–111) Recover for 40 seconds	Arrows for 20 seconds (see pp. 114–115) Recover for 40 seconds

Repeat AMRAP for the time remaining of the 20 minutes.

DAY SEVEN

Big Jump Jacks for 20 seconds (see pp. 106–107) Recover for 40 seconds	Thrusters for 20 seconds (see pp. 98–99) Recover for 40 seconds	Gecko for 20 seconds (see pp. 100–101) Recover for 40 seconds	Bangbang Firing for 20 seconds (see pp. 110–111) Recover for 40 seconds	Arrows for 20 seconds (see pp. 114–115) Recover for 40 seconds

Repeat AMRAP for the time remaining of the 20 minutes.

DAY EIGHT

Big Jump Jacks for 20 seconds (see pp. 106–107) Recover for 40 seconds	Thrusters for 20 seconds (see pp. 98–99) Recover for 40 seconds	Gecko for 20 seconds (see pp. 100–101) Recover for 40 seconds	Bangbang Firing for 20 seconds (see pp. 110–111) Recover for 40 seconds	Arrows for 20 seconds (see pp. 114–115) Recover for 40 seconds

Repeat AMRAP for the time remaining of the 20 minutes.

DAY NINE

Rest day from HIIT but not from all activity. Stretch, do mobility exercises and be as active as you can without actually 'working out'.

DAY TEN

BurPEE for 30 seconds *(see pp. 112–113)* Recover for 60 seconds	Plunges for 30 seconds *(see pp. 104–105)* Recover for 60 seconds	Run-ups for 30 seconds *(see pp. 102–103)* Recover for 60 seconds	Gecko for 30 seconds *(see pp. 100–101)* Recover for 60 seconds	Skater for 30 seconds *(see pp. 116–117)* Recover for 60 seconds

Repeat AMRAP for the time remaining of the 20 minutes.

DAY ELEVEN

BurPEE for	Plunges for	Run-ups for	Gecko for	Skater for
30 seconds	30 seconds	30 seconds	30 seconds	30 seconds
(see pp. 112–113)	(see pp. 104–105)	(see pp. 102–103)	(see pp. 100–101)	(see pp. 116–117)
Recover for	Recover for	Recover for	Recover for	Recover for
60 seconds	60 seconds	60 seconds	60 seconds	60 seconds

Repeat AMRAP for the time remaining of the 20 minutes.

DAY TWELVE

BurPEE for	Plunges for	Run-ups for	Gecko for	Skater for
30 seconds	30 seconds	30 seconds	30 seconds	30 seconds
(see pp. 112–113)	(see pp. 104–105)	(see pp. 102–103)	(see pp. 100–101)	(see pp. 116–117)
Recover for	Recover for	Recover for	Recover for	Recover for
60 seconds	60 seconds	60 seconds	60 seconds	60 seconds

Repeat AMRAP for the time remaining of the 20 minutes.

DAY THIRTEEN

BurPEE for	Plunges for	Run-ups for	Gecko for	Skater for
30 seconds	30 seconds	30 seconds	30 seconds	30 seconds
(see pp. 112–113)	(see pp. 104–105)	(see pp. 102–103)	(see pp. 100–101)	(see pp. 116–117)
Recover for	Recover for	Recover for	Recover for	Recover for
60 seconds	60 seconds	60 seconds	60 seconds	60 seconds

Repeat AMRAP for the time remaining of the 20 minutes.

DAY FOURTEEN

Rest day from HIIT but not from all activity. Stretch, do mobility exercises and be as active as you can without actually 'working out'.

DAY FIFTEEN

Do the HIIT Test
Compare your score to the start of the challenge, any improvement in such a short period of time is good. You are now past the half-way point. Revel in your improved performance and see if you can push yourself further for the second half of the programme.

DAY SIXTEEN

Arrows for
30 seconds
(see pp. 114–115)
Recover for
60 seconds

Bangbang
Firing for
30 seconds
(see pp. 110–111)
Recover for
60 seconds

Rabbit for
30 seconds
(see pp. 108–109)
Recover for
60 seconds

Big Jump Jacks
for 30 seconds
(see pp. 106–107)
Recover for
60 seconds

Thrusters for
30 seconds
(see pp. 98–99)
Recover for
60 seconds

Repeat AMRAP for the time remaining of the 20 minutes.

DAY SEVENTEEN

Arrows for
30 seconds
(see pp. 114–115)
Recover for
60 seconds

Bangbang
Firing for
30 seconds
(see pp. 110–111)
Recover for
60 seconds

Rabbit for
30 seconds
(see pp. 108–109)
Recover for
60 seconds

Big Jump Jacks
for 30 seconds
(see pp. 106–107)
Recover for
60 seconds

Thrusters for
30 seconds
(see pp. 98–99)
Recover for
60 seconds

Repeat AMRAP for the time remaining of the 20 minutes.

DAY EIGHTEEN

Arrows for
30 seconds
(see pp. 114–115)
Recover for
60 seconds

Bangbang
Firing for
30 seconds
(see pp. 110–111)
Recover for
60 seconds

Rabbit for
30 seconds
(see pp. 108–109)
Recover for
60 seconds

Big Jump Jacks
for 30 seconds
(see pp. 106–107)
Recover for
60 seconds

Thrusters for
30 seconds
(see pp. 98–99)
Recover for
60 seconds

Repeat AMRAP for the time remaining of the 20 minutes.

DAY NINETEEN

Rest day from HIIT but not from all activity. Stretch, do mobility exercises and be as active as you can without actually 'working out'.

DAY TWENTY

Do the HIIT Test
Compare your score to the start of the challenge, any improvement in such a short period of time is good. You are now two-thirds of the way through the programme. Push on, you're on the home straight!

DAY TWENTY-ONE

Run-ups for
30 seconds
(see pp. 102–103)
Recover for
60 seconds

Gecko for
30 seconds
(see pp. 100–101)
Recover for
60 seconds

Run-ups for
30 seconds
(see pp. 102–103)
Recover for
60 seconds

Skater for
30 seconds
(see pp. 116–117)
Recover for
60 seconds

BurPEE for
30 seconds
(see pp. 112–113)
Recover for
60 seconds

Repeat AMRAP for the time remaining of the 20 minutes.

DAY TWENTY-TWO

Run-ups for
30 seconds
(see pp. 102–103)
Recover for
60 seconds

Gecko for
30 seconds
(see pp. 100–101)
Recover for
60 seconds

Run-ups for
30 seconds
(see pp. 102–103)
Recover for
60 seconds

Skater for
30 seconds
(see pp. 116–117)
Recover for
60 seconds

BurPEE for
30 seconds
(see pp. 112–113)
Recover for
60 seconds

Repeat AMRAP for the time remaining of the 20 minutes.

DAY TWENTY-THREE

Run-ups for
30 seconds
(see pp. 102–103)
Recover for
60 seconds

Gecko for
30 seconds
(see pp. 100–101)
Recover for
60 seconds

Run-ups for
30 seconds
(see pp. 102–103)
Recover for
60 seconds

Skater for
30 seconds
(see pp. 116–117)
Recover for
60 seconds

BurPEE for
30 seconds
(see pp. 112–113)
Recover for
60 seconds

Repeat AMRAP for the time remaining of the 20 minutes.

DAY TWENTY-FOUR

Rest day from HIIT but not from all activity. Stretch, do mobility exercises and be as active as you can without actually 'working out'.

DAY TWENTY-FIVE

Run-ups for	Arrows for	Run-ups for	Rabbit for	Thrusters for
45 seconds	45 seconds	45 seconds	45 seconds	45 seconds
(see pp. 102–103)	*(see pp. 114–115)*	*(see pp. 102–103)*	*(see pp. 108–109)*	*(see pp. 98–99)*
Recover for	Recover for	Recover for	Recover for	Recover for
45 seconds	45 seconds	45 seconds	45 seconds	45 seconds

Repeat AMRAP for the time remaining of the 15 minutes.

DAY TWENTY-SIX

Run-ups for	Arrows for	Run-ups for	Rabbit for	Thrusters for
45 seconds	45 seconds	45 seconds	45 seconds	45 seconds
(see pp. 102–103)	*(see pp. 114–115)*	*(see pp. 102–103)*	*(see pp. 108–109)*	*(see pp. 98–99)*
Recover for	Recover for	Recover for	Recover for	Recover for
45 seconds	45 seconds	45 seconds	45 seconds	45 seconds

Repeat AMRAP for the time remaining of the 15 minutes.

DAY TWENTY-SEVEN

Run-ups for 45 seconds *(see pp. 102–103)*	Thrusters for 45 seconds *(see pp. 98–99)*	Run-ups for 45 seconds *(see pp. 102–103)*	Bangbang Firing for 45 seconds *(see pp. 110–111)*	BurPEE for 45 seconds *(see pp. 112–113)*
Recover for 45 seconds	Recover for 45 seconds	Recover for 45 seconds	Recover for 45 seconds	Recover for 45 seconds

Repeat AMRAP for the time remaining of the 15 minutes.

DAY TWENTY-EIGHT

Do the HIIT Test

Compare your score to the start of the challenge, any improvement in such a short period of time is good, however, anything greater than a 10–15% improvement is awesome. Take an 'after' selfie and just see how good you look compared to the 'before' selfies taken 28 days ago.

Now what? You could just continue to eat, sleep, and repeat the 28-Day 'Zero to Super Hero' challenge but if I was you I would be rewarding myself with the purchase of some dumbbells or a TRX Suspension Trainer to enable you to enjoy and add variety to your HIIT sessions (see pp. 144–145).

PART SIX

QUICK-HIIT ANSWERS

Is HIIT just a trend that's here today and gone tomorrow or is it here to stay?

Yes, I believe it is very much here to stay and actually the time is right for it to be a big hit (excuse the pun). Unlike core training and functional exercise, it's a much easier thing to get your head around because the benefits seem to fairly accurately reflect the amount of effort you put into the process. Often, you find with new trends in the world of fitness that people drop other stuff for a while then go back to it. HIIT, however, complements everything and competes with nothing, so in reality it's something that you can do 52 weeks of the year as your main form of exercise, or perhaps just as a supplement to other styles of activity when you have less time available or are working towards specific goals.

How does HIIT work?

In its rawest state, HIIT is just a bunch of numbers that relate to the level of intensity you perform at and the amount of time for which you work, compared to the amount of time for which you recover before doing the work again. This simple format has led to the development by scientists and big businesses of a huge range of formats that can be classed as HIIT. These include anything from programmes that prescribe just seconds of activity per week up to far more traditional 30-minute sessions that are delivered in a group-exercise format in health clubs and gyms under brand names such as Les Mills Grit™, INSANITY® and Tabata™.

Depending on your age, level of fitness, weight and medical status it's true to say there will be a style of HIIT for you. At one end of the athletic spectrum it has been tested and used on extremely overweight diabetics in a controlled environment, while at the other end some of the fittest bodies on the planet are the result of HIIT activity (so long as you are in the group of people who think a sprinter's muscular, athletic body is more aspirational than the very slim physique of a marathon runner).

Ok, health is great but is HIIT going to change the way I look on the beach?

Yes, so long as you don't think HIIT is an excuse to eat, drink and be merry 24/7 because this magic exercise format will right all those wrongs, and provided that looking 'good' to you means reducing the levels of fat that cover your muscles. It's also good to know that while all of the aesthetic improvements are happening you'll also be improving your body's sensitivity to insulin (which means your body copes better with an influx of sugar) and strengthening your heart and lungs.

I did HIIT and actually threw up – is that normal?

No, this shouldn't be normal. You probably just need to sort out your pre-workout eating habits. Avoid eating anything one hour before your session, in particular heavy, slow-to-digest foods as eating these can have the knock-on effect of redirecting lots of blood that should be in your muscles for the HIIT session to your guts as part of the digestive process.

I'm pregnant – to HIIT or not to HIIT?

Firstly, congratulations! The answer depends on whether you were HIITing before you became pregnant, how advanced you are and, if we are honest, how big you are. Obviously as a man, any advice is based on observations made when working with and around pregnant women. If you were not doing regular HIIT sessions for the six-month period before conception, then NO I don't think it is wise to begin now. If you were then there are still 'buts' (and you must speak to your GP before doing any exercise). Questions that may be asked are:

☐ Is this your first pregnancy (this is relevant because if not then we have no precedent to refer to);

☐ Have there been any issues, such as raised blood pressure or conditions that are exacerbated by exercise and especially high-intensity exercise?

Even if the answer to these is no, please don't take my word for it and push yourself hard. Please visit your GP or healthcare practitioner who can offer further advice.

Can I train for a marathon doing just HIIT?

Yes, and no … I know of people who have run multiple marathons in under three hours with far less actual running than is promoted by conventional training programmes, by using HIIT to condition themselves and make them resilient (resistant) to being stressed by activity. This involves doing HIIT with bodyweight and weights up to five times per week and then in effect skipping the 'long run' that most people training for marathons aim to do at least once a week in the build-up to a marathon. However, I think that running, swimming and cycling all rely upon an element of skill development as well as fitness – there is a lot of skill involved in pacing yourself and maintaining your posture and technique throughout a marathon. In my opinion, therefore, you can use HIIT as part of your training programme but you need to also do the long runs to establish the required mental resilience to hold form, and the muscle memory to keep delivering the skillful strides that will get you around fast.

Is HIIT better if I use equipment?

Yes, and no. The joy of HIIT is that if you're inclined to do so you could do a session in a field, gym, hotel room, kitchen, garage or basically anywhere that has a good, safe surface to work on and sufficient space for you to be able to move (including overhead height). What using equipment adds to the entire experience is: 1) a wider range of potential exercises; and 2) the ability to add intensity without increasing the speed of movement. If I ask you to do a fast squat then in order to get your heart pumping without a weight you need to attack the tempo; if you were doing exactly the same movement holding a weight (medicine ball, dumbbell, kettlebell) in your hands then the speed could be reduced. You'd achieve the same outcome either way. The short answer is that if you have some equipment then so long as using it improves the outcome then go for it, but it's absolutely not necessary to have equipment to do HIIT.

Can I do HIIT every day?

No, not if you do it properly. HIIT 'hurts' while you are doing it, and in a good way it 'damages' muscle fibres and stresses the tissues within our respiratory system, which all takes time to repair. Doing HIIT three times per week with a day in between each session is a good plan of action. The buffer/recovery day doesn't have to be a day of inactivity – you just shouldn't do exercise that challenges the muscles and system in the way HIIT does. So, for instance, doing strength training or mobility/flexibility sessions is the perfect way to complement the HIIT and maximise the returns.

Help, I'm scared. Are you sure this isn't just for elite athletes?

In the same way that we can use various physiological markers to measure intensity, we can use psychological ones to help you achieve your maximum potential, and in particular one called 'perceived exertion'. This involves you asking and answering the question: 'How hard am I working compared to how hard I think I could work?' This is a constantly moving goalpost because as a beginner what you think of as your 10 out of 10 today won't be the same after a few weeks or months of regular activity. As for these athletes, well yes, there is an element of nature involved. That said, there is also a lot of nurture, meaning that they didn't start out that good and have spent many years getting to their current level of ability. The same mindset nonetheless applies: their 10 out of 10 is very different now from what it was during earlier stages of their development.

Provided that you have 'passed' the reality check (see p. 13) then you shouldn't be scared. Just be sensible and listen to your body, accept where you are today in terms of ability, and remember that if you do nothing then your level of ability will probably get worse, whereas if you start training now the most likely outcome is that you will start to feel better within two or three weeks, and other people will actually start to notice the difference in six to eight weeks. How cool is that!

Can I do HIIT if I am injured or feeling ill?

Depends. Some problems such as a wrist or hand injury can probably be accommodated without any knock-on effects, but an existing injury in the shoulder, spine, hips or lower body will be harder to work around. Most injuries require a careful balance of work and recovery so you need to treat each one on a case-by-case basis. It is important to listen to your body and, when in doubt, consult a fitness specialist for their expert opinion. This takes precedence over any training programme outlined here.

There are two types of exercise: 'Energy In' which is the meditation styles of exercise such as are yoga and Pilates, and 'Energy Out' which is the high-exertion type that makes you sweat, out of breath and result in fatigue. HIIT is definitely 'Energy Out' along with running, swimming, weightlifting and most other sports. If you are feeling unwell then any form of 'Energy Out' exercise will make you feel worse and prolong the illness. With the exception of very light colds, if you have a fever, congestion and/ or a runny nose then take a break from exercise until you feel better.

Should I stretch during my warm-up?

Not if when you say stretch you mean static stretching ... this isn't 1985! Doing static stretches as part of getting ready for a workout is a rather dated approach, and in fact there have been many athletes who have realised that when they stopped doing static stretches their performance improved, especially if their chosen activity demanded strength. Static stretches might feel nice but they have the physiological and neurological effect of reducing the individual muscles fibres' ability to perform maximal contractions.

Mobility exercises, however, should absolutely be performed before HIIT sessions, as should dynamic stretches. Both of these have the benefit of complementing the requirements of muscles used during HIIT by firing them up rather than damping them down. There is warm-up and cool-down advice in Part 4 of the book.

Can I do HIIT in a swimming pool?

Only if your technique is good enough for you to be able to work hard in a productive way. If your standard of swimming is 'recreational' rather than 'athletic' then you most likely won't generate the kind of intensity that will challenge your body in the right way. If you can sprint for 50m or more doing freestyle, including tumble turns and breathing on both sides, then the chances are you are good enough. The only caveat I would add is that maintaining good technique is critical when swimming so pushing yourself to exhaustion in the pool with HIIT sessions may a have an undesirable knock-on effect to the quality of your stroke.

Should I stretch before and after a HIIT workout?

In recent years pre- and post-workout stretches have virtually disappeared from training programmes. This could be down to time constraints, but is most likely because people are confused about the benefits. The thinking was that these stretches reduced the risk of injury and helped reduce post-exercise muscle pain known as DOMS (delayed onset muscle soreness). We now believe this to be unlikely, as DOMS is thought to be the result of damage/tears to the muscle fibres that have been recently worked. Workouts that include pre-exercise mobility exercises (rather than traditional static stretches) are, to me, the most productive.

Similarly for post-workout stretches, I don't feel any benefit from doing static stretches. I do however find it beneficial to spend a couple of minutes resetting my body to its pre-exercise posture. Depending on which body zone the workout targeted, I'll move through each area of the body that feels tight, moving the area in big sweeping movements until it feels as if the muscles have released.

Cooling down to 'flush out' waste products from muscles and the bloodstream is now considered to be unproductive. The original thinking was that we got sore on account that there was lactic acid 'trapped' in the muscle. This is now considered to be a myth due to lactic acid in fact being continuously metabolised rather than 'left behind'.

AUTHOR BIOGRAPHY

STEVE BARRETT is a leading international fitness presenter who has appeared multiple times at the 'Big Five' global fitness events which include FitPro (UK), IDEA (USA), IHRSA (USA), Filex (Australia) and MIOFF (Russia).

His career in the fitness industry as a personal trainer spans over 25 years. His work as a lecturer and presenter has taken him to 55 countries including the United States, Russia and Australia. He is a former national competitor in athletics, rugby, mountain biking and sport aerobics.

For many years Steve delivered Reebok International's fitness strategy and implementation via their training faculty Reebok University. He gained the title of Reebok Global Master Trainer, which is a certification that required a minimum of three years' studying, presenting and researching both practical and academic subjects. As his career evolved he progressed from being a Master Trainer to the coveted role of Global Brand Ambassador to some of the worlds biggest fitness equipment manufacturer's including Matrix and Escape Fitness.

Steve played a key role in the development of the training systems and launch of two significant products in the fitness industry: the Reebok Deck and Reebok Core Board®. As a personal trainer, in addition to teaching the teachers and working with the rich and famous, he has been involved in the training of many international athletes and sports personalities at Liverpool FC, Arsenal FC, Manchester United FC, the Welsh RFU, and UK athletics. Within the fitness industry he has acted as a consultant to leading brand names, including Nestlé, Kelloggs, Reebok and Adidas. Steve developed the original exercise programmes for the Kelloggs' Special K two-week 'Drop A Dress Size' programme that was watched and downloaded by millions of women worldwide. His media experience includes being guest expert for the BBC and writing for numerous publications including *The Times*, *The Independent*, *The Daily Telegraph*, *Runner's World*, *Men's Fitness*, *Rugby News*, *Health & Fitness*, *Zest*, *Ultra-FIT*, *Men's Health* UK and Australia, and many more.

Steve is the author of the celebrated *Total Workouts* books, with each title highlighting the trade secrets of a personal trainer on the subject of strength and conditioning for foam rolling, dumbbells, suspended bodyweight systems, kettlebells and gym balls. Steve's expertise is in the development of logical, user friendly, safe and effective training programmes. The work that he is most proud of, however, isn't his celebrity projects, but the changes to ordinary people's lives that never get reported.

Now that he has been teaching fitness throughout his 20s, 30s and now 40s, he has developed a tremendous ability to relate to the challenges that people face to incorporate exercise into their lifestyle, and while the fitness industry expects personal trainers to work with clients for a short period of time, Steve has been working with many of his clients for nearly two decades, continuously evolving to meet their changing needs. His fun and direct approach has resulted in many couch potatoes running out of excuses and transforming into fitness converts.

INDEX

14-Day 'Zero to Hero' Challenge 208–213
28-Day 'Zero to Super Hero' Challenge 214–225

A
aerobic system 31
aesthetic improvements 229
after-burn see Excess Post-Exercise Oxygen Consumption (EPOC)
ageing process 25–26, 88
alcohol 87, 89
American College of Sports Medicine (ACSM) 16, 29, 30, 59
American Council on Exercise (ACE) 59
anaerobic system 31
anaerobic threshold 24
Apps 67
Arrows 114–115
assessment 202–4
    FastLap® test 203
    HIIT Test 202, 203–204
    Personal Best (PB) 202
ATP-PC system 31

B
Bangbang Firing 110–111
barbells 71
Barry's Bootcamp™ 33, 57
basal metabolic rate (BMR) 84
Bell Ringer 126–127
benefits of HIIT 20–27
    ageing, slowing 25–26
    blood quality 24
    brain function 24–25
    cardiovascular fitness 21–22
    cortisol levels 26–27
    endurance levels 22
    human growth hormone 22–23
    mood enhancement 26
    weight loss 21
Big Jump Jacks 106–107
blood glucose levels 22, 193

blood quality 24
body fat 13, 21, 26, 31, 45, 50, 88
bodyweight exercises 96–117
    with dumbbells 118–139
bodyweight workouts 173–182
    bodyweight moves (no equipment) 177–178
    bodyweight moves (with dumbbells) 179–182
    intensity modes 174–176
Bottom Up 120–121
Box Jumps 150–151
brain function 24–25, 85
brain-derived neurotrophic factor (BDNF) 24
brand names 32–57
bras, sports 66
BurpEE 112–113
'buzz' 15, 24, 31

C
caffeine 89
calories
    burning 21, 57, 160
    restricted-calorie diet 13
carbohydrates 85, 93
cardiovascular fitness 21–22, 193
    cardio sessions on machines 152–165
cholesterol 10, 24, 50
circadian rhythm 23
clothing 64–66
coffee 89
cooling down 170, 235
core strength 114, 144, 147, 173
cortisol 26–27, 93
creatine phosphate (CP) 31
CrossFit™ 33, 40–41
cycling 197–199
    base training 197
    HIIT sessions 197, 199
    sprint efforts 197

D
dehydration 8, 13, 89
Delayed Onset Muscle Soreness (DOMS) 9, 89, 235
diabetes 44, 88, 228
diet
    restricted-calorie diet 13
    see also nutrition
dizziness 10, 12
Double Thruster 124–125
dumbbells
    bodyweight exercises 118–139
    weights 179
    workouts with 179–182
dynamic stretches 234

E
electrolyte tablets 89
endocrine system 22
endorphins 24, 26
endurance 22, 89
'Energy In' exercise 234
'Energy Out' exercise 234
equipment, gym-based 68–81, 231
    Olympic bar and weight discs 71
    plyo box 73
    punch bag 81
    rig 70
    rowing machine 78–79
    running machine 74
    sledge 80
    spin bike 75–77
    TRX 69
estimated maximum heart rate (EMHR) 30
euphoria, post-exercise 24, 31
Excess Post-Exercise Oxygen Consumption (EPOC) 21, 31, 93
exercise monitors 67

F
FastLap® test 203
fats 85–86
fitness

baseline level of 61
  participant screening 59–60
  self-assessment 61
foam roller 170
footwear 65
fructose 23, 88

G
gadgetry 67
Gecko 100–101
glycogen 31, 45, 89, 93
gym-based exercises 140–151
gym-based workouts 183–189
  intensity modes 185

H
Halo 138–139
Hang Tough Combo 148–149
health checklist 12
HIIT
  assessment 202–204
  brand names 32–57
  effectiveness 16, 17
  equipment 17
  FAQs 228–235
  kit 64–81
  overview 15–17, 228
  participant screening 59–60
  reality check 13, 232
  research evidence 21, 23, 24, 25, 26, 44–45, 50
  reservations about 16, 17, 59
  ROI (return on investments) 20–27
  safety 8–11, 59–61
  science of 28–31
  sessions, frequency of 232
  Training Challenges 205–225
  versions of 15, 17, 228
HIIT Test 202, 203–204
human growth hormone (HGH) 22–23, 50, 51
  synthetic 23
hydration 87, 89–90

hydrogenated fats 25, 86
hyponatremia 90

I
injuries 8, 60, 80, 193
  working out with 234
INSANITY® 33, 36–37
insulin
  sensitivity 229
  sugars and 23
Integrated Fitness Training (ITF) model 59
International Dance-Exercise Association (IDEA) 59
International Health, Racquet and Sportsclub Association (IHRSA) 59
intervals 15
  duration 30

K
kettlebells 180
kit 64–81
  clothing 64–66
  gadgetry 67
  gym-based equipment 68–81

L
lactic acid 235
Les Mills Grit™ 33, 46–47
limits, understanding your 9
listening to your body 9, 10
The Little Method 33, 44–45

M
macronutrients 83–86
marathon running 22, 231
Matrix Fitness 50, 51
maximum oxygen consumption (VO2 max) 22
medical advice 10–11, 12, 230
medical history, personal 10, 12
metabolism 22, 44, 45, 84
micronutrients 86

mitochondrial biogenesis 45
mobility exercises 169, 232, 234, 235
monounsaturated fats 86
mood enhancement 26
muscles
  age-related muscle decay 24
  Delayed Onset Muscle Soreness (DOMS) 9, 89, 235
  fast-twitch fibres 23
  metabolic capacity 44–45
  'muscle confusion' 43
  muscle isolation 27, 70
  muscle tissue development 22, 25
MYZONE® chest strap 67

N
National Exercise Trainer Association (NETA) 59
neuroplasticity 24
neurotrophic factors (NF) 25
nutrition 82–93
  'Golden Rules' 87
  hydration 87, 89–90
  macronutrients 83–86
  micronutrients 86
  post-workout nutrition 91, 93
  pre-workout nutrition 90–91
  vegetarians 83, 86

O
obesity 10, 25, 50
Olympic bar and weight discs 71
Orangetheory Fitness 33, 55
outdoor workouts 192–199
oxygen absorption 21, 22, 24, 31

P
P90X™ 33, 42–43
participant screening 59–60
perceived exertion 232
Peripheral Heart Action (PHA) 43
Personal Best (PB) 202

pituitary gland 22
Plunges 104–105
plyo box 73
    Box Jumps 150–151
polyunsaturated fats 86
pregnant women 230
primal movement patterns 35
protein 84, 90, 93
protein powder 84
prowler see sledge
punch bag 81

R
Rabbit 108–109
ramp sessions 44
reality check 13, 232
rebound jumping 73
recovery days 206, 232
recovery periods 15, 30
    duration 30
research evidence 21, 23, 24, 25,
    26, 44–45, 50
rig 70
rowing machine 78–79
    medium resistance/fast strokes
    164–165
    using 79
Run-ups 102–103
'runner's high' 24
running 192–196
    HIIT sessions 194–195, 196
    inclines 196
    marathons 22, 231
    off-road 195–196
    recovery periods 193, 194–195
    speed work 193
    on the spot 102–103
    sprinting 193, 196
    on the track 194–195
running machine 74
    hike mode 158–159
    push mode 156–157
    run mode 154–155
    sledge mode 157
    using 74

S
safety 8–11, 59–61
saturated fats 86
sessions, frequency of 232
Shot-put Driver 134–135
Skater 116–117
sledge 80
sleep 23, 93
smoking 10, 25, 88
snacks 87, 91
somatopause 23
SoulCycle™ 33, 54, 75
space requirement 17, 173
spin bike 75–77
    features 76
    resistance 76, 160, 163
    speed mode 162–163
    standard mode 160–161
    using 77
Split and Rip 122–123
spring interval training method 30
Sprint 8™ 33, 50–51
sprinting 193, 196
    cycle sprints 197
    on running machine 154
Squat Rack Drop and Row Combo
    142–143
static stretches 170, 234, 235
Stones Lift 130–131
stretching 170, 234, 235
sugar 23, 88
Sumo Squat - Shoulder Press
    Combo 146–147
suspension trainer see TRX
swimming, HIIT and 235
Swing and Step 128–129

T
Tabata™ 30, 33, 34–35, 160
    Tabata™ protocol 35, 44, 45
telomere decay 25–26
throwing up 229
Thrusters 98–99
time commitment 17, 20, 173
timing sessions 67

trainers/sneakers 65
Training Challenges 205–225
    14-Day 'Zero to Hero' Challenge
    208–213
    28-Day 'Zero to Super Hero'
    Challenge 214–225
trans fats 25, 86
treadmill see running machine
tri-planar movements 26, 27, 135,
    139
TRX 69
TRX Pike Pull and Y-fly 144–145
TRX Skydive 146–147
Turbulence Training 33, 48
Turk Thruster 136–137

U
unfit people 60, 228

V
V-jump 132–133
venous return system 24
vitamin supplements 86
VO2 max see maximum oxygen
    consumption

W
walking 173
warming up 169–170, 234, 235
water, drinking 87, 89–90
weight loss 21
work/rest ratios 30, 31
workouts 166–170
    bodyweight moves 173–182
    cooling down 170, 235
    gym-based 183–189
    outdoor 192–199
    time commitment 17, 20, 173
    warming up 169–170, 234, 235